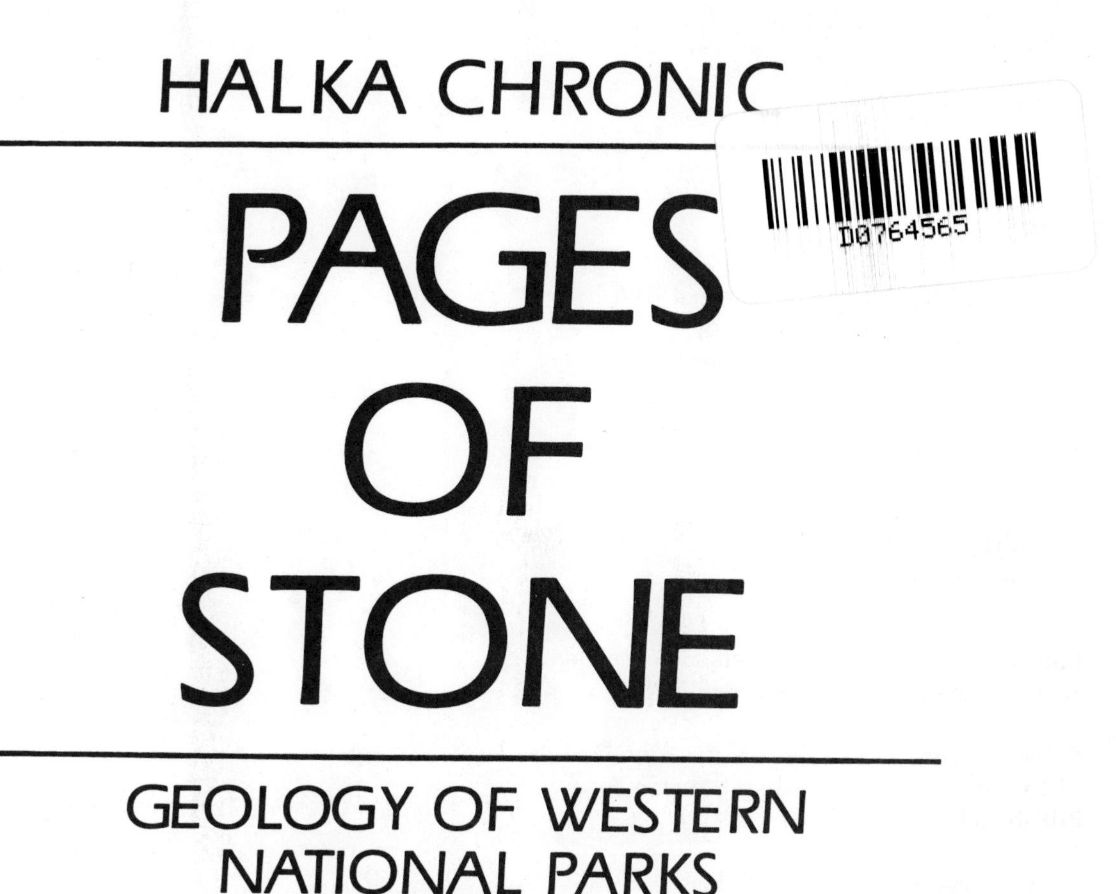

HALKA CHRONIC

D0764565

PAGES OF STONE

GEOLOGY OF WESTERN NATIONAL PARKS & MONUMENTS

2: SIERRA NEVADA, CASCADES, & PACIFIC COAST

The Mountaineers · Seattle

THE MOUNTAINEERS: Organized 1906
"... to explore, study, and enjoy
the natural beauty of the outdoors."

© 1986 Halka Chronic
All rights reserved

Published by The Mountaineers, 306 Second
Avenue West, Seattle, Washington 98119

Published simultaneously in Canada by Douglas &
McIntyre, Ltd., 1615 Venables Street, Vancouver,
British Columbia V5L 2H1

Cover design by Constance Bollen
On the cover: Half Dome, Yosemite National Park.
Inset: Jointed volcanic rock, Mount Rainier National Park.

Manufactured in the United States of America

Carl Sagan quote on page iii reprinted with permission of Dr. Sagan,
from *Broca's Brain,* Random House, New York, 1979.

0 9 8 7 6
5 4 3 2 1

Library of Congress Cataloging-in-Publication Data
(Revised for vol. 2)

Chronic, Halka.
Pages of stone.

Includes bibliographies and indexes.
Contents: v. 1. Rocky Mountains and Western Great
Plains — v. 2. Sierra Nevada, Cascades, and Pacific
Coast.
1. Geology—West (U.S.) 2. National parks and
reserves—West (U.S.) 3. Natural monuments—West (U.S.)
QE79.C47 1984 557.3 82-422
ISBN 0-89886-095-4 (pbk. : v. 1)
ISBN 0-89886-114-4 (pbk. : v. 2)

Science is a joy. It is not just something for an isolated, remote elite. It is our birthright.

Carl Sagan

Acknowledgments

This book is designed primarily for the reading public, for those without a formal geologic education. I hope that it will prove interesting and informative to students and professionals as well. Its information comes from the geologic literature and from individual geologists who have worked in and near the park areas, as well as from personal observations. Park Service naturalists have been unstinting in their cooperation. In particular I wish to express my gratitude to Dwight R. Crandall, Charles Baker, Edwin H. McKee, and Don Peterson of the U.S. Geological Survey, David D. Alt of the University of Montana, and Vincent Matthews III of Lear Petroleum Exploration, Inc., all of whom critiqued portions of the manuscript. Unless otherwise indicated, the photographs are my own.

Contents

Introduction 1

Part 1 Our Geologic Legacy

I. Continents That Drift 3
II. Rocks, Time, and Fossils 9
III. The Making of Mountains 16
IV. The Wearing Away 21
V. Understanding Maps and Diagrams 27
VI. Other Reading 28

Part 2 National Parks and Monuments

Cabrillo National Monument 31
Channel Islands National Park 35
Crater Lake National Park 40
Devils Postpile National Monument 49
John Day Fossil Beds National Monument 54
Lassen Volcanic National Park 59
Lava Beds National Monument 71
Mount Rainier National Park 75
Mount St. Helens National Volcanic Monument 90
North Cascades National Park 100
Olympic National Park 108
Oregon Caves National Monument 121
Pinnacles National Monument 125
Redwood National Park 129
Sequoia and Kings Canyon National Parks 135
Yosemite National Park 146

Glossary 159
Index 166

Wave-worn pebbles of volcanic rock on a Pacific beach are clues to the nature of the Olympic Mountains.

Introduction

The Earth is a history book, vividly written, copiously illustrated. We take it from its shelf, open its dusty pages, and read its story. Its chapters tell of ancient seas and ancient lands, of arid deserts and steaming jungles, of rivers and lakes and floods and droughts, of bursting volcanoes and far-flowing rivers of lava. With fascination we read of molten rock deep beneath its cooling crust, of mountains formed and then eroded away, of life developing in long-gone seas, emerging eventually upon the land and there maintaining itself despite the fearful elements of sea and sky.

Courtesy of NASA

Though many pages of the Earth's history book are missing, here and there are fascinating chapters that vividly portray the story of our planet. Piecing together torn pages, we learn gradually how the planet we live on is put together, a Chinese puzzle that, assembled, equals a hot, partly fluid ball, thinly crusted, held together by its own gravity, sailing with its seas and continents serenely through the immensity of space.

By human standards, the Earth is enormously old—so enormously old that we can hardly conceive of its age in a sensible way. But if we put together an analogy, one that likens the 4.6 billion years of the Earth's existence to 46 years—scarcely two-thirds of a human lifetime, we find it easier to understand:

If the Earth is 46 years old, human beings—members of the genus *Homo*—have lived upon it for only eleven days. During the last twelve minutes or so these inquisitive two-footed creatures, able to use their hands for purposes other than walking and running, have tried to keep track of their own history. In the last two minutes they have made some strong attempts to decipher the long, long story of the planet they live on. Yet it was only eight seconds ago that they developed a systematic understanding of its origin and the changes that have taken place upon and within it.

Translated to real time, the planet is, as I said above, about 4.6 billion years old. Man has walked its surface for something like 3 million years. About 2450 years ago Herodotus gave us the first history book—a chronicle of political and military struggles between Persia and Greece. A little over 400 years ago geology as a science was born when Georgius Agricola published his observations of mineral deposits in Bohemia, now western Czechoslovakia. And a mere 25 years ago the Plate Tectonic Theory was born, a theory that elegantly interprets the origins and migrations of the present continents.

Unraveling the geologic story, putting together the torn pages of the Earth's great book, geologists search their outdoor laboratory for clues to the past. They hike the hills, rock hammer in hand, to map and measure changes in rock types. They fly above the land or dive beneath the sea. They collect specimens, photograph landforms, examine small features within the rocks: diagnostic minerals, tiny fracture patterns, distinctive fossils. Back in indoor laboratories they employ tools not available to geologists a century or even half a century ago: electron microscopes, scintillometers, mass spectrometers, and delicate chemical analyses. With a battery of techniques they determine the age of their rocks, their probable origins. And with other tools they draw further secrets: clues to the Earth's past magnetism and to collisions in space with asteroids or other interplanetary bodies.

As geologists put their clues together and examine them in the light of clues collected by their predecessors and contemporaries, they come up with geologic maps that show the kinds of rocks exposed at the surface, whether they are flat-lying or tilted, and the position of folds and faults that disrupt expected patterns. Because the record of ancient animal life is also there in the rocks, some of their work and many of their clues converge with the work of other scientists—biologists. Together the two groups of scientists learn of the origins of modern animal and plant species and record the way that living things cope with an ever-changing world. Fossils preserved in rock help to date that rock, and in addition furnish undeniable evidence of the theory of evolution: that the animal and plant species around us today evolved through time by a process of natural selection or "survival of the fittest."

Thus have geologists worked, spanning missing pages with data from chapters more intact. For geology is, like all science, logical—a progression of ideas based on previous findings.

Much of what we know about our planet we learn from things about us—things that can be looked at, touched, recorded, analyzed. Geology unlocks the history of the Earth by using the present as the key to the past. As portions of the Earth's crust rise and fall today, or break along great belts of faults, with each movement perceptibly or imperceptibly shaking the land, so past mountains have been raised again and again. Within human lifespans, old volcanoes have awakened to belch fiery molten lava, to create new mountains or to reshape old ones, just as they have done repeatedly during the history of the Earth. Slowly the mountains are worn down, as were mountains of the past. Today's streams and rivers wash rocks and sand and silt from today's highlands and deposit them on floodplains and deltas far from their source, just as did rivers of the past. The debris of today's continents, sand, clay, and limy mud, comes to rest in horizontal layers on the sea floor, as it did in the past, forming layered strata that will harden into sandstone, shale, and limestone. And in the strata, present and past, is buried a record of life—the shells and body parts of animals and plants, parts of the age-old geologic puzzle.

Some of the most alluring chapters in the history book of the Earth we find in our national parks and monuments, where scenic beauty is born of, and reflects, the geologic story. Taken together, the parks and monuments illustrate most of the principles on which the study of geology is based. In them are rocks formed from molten magma, volcanoes new and old, alive and dead. In them are illustrations of sedimentary processes by which new rock is formed from old, by wind or running water or glacial ice. In them are rocks altered by high temperatures and great pressures, examples of processes going on even today deep in the crust of the Earth. In the park areas—in both parks and monuments—also are examples of faults and folds and other geologic structures, and a multitude of displays of the opposing forces of uplift and erosion. In visiting our national parks and monuments you will, I hope, come to understand not only these splendid chapters of the Earth's history, but the more prosaic pages between.

National park areas are crisscrossed with inviting roads and trails. Whether you travel on foot or by car (I hope you'll do both), allow time to stop and examine rocks by the wayside. Look at landforms—mountains, plains, valleys—and ponder their origins. Park viewpoints may help you with explanatory displays. Guide leaflets will lead you on many trails and along some highways. At visitor centers you can find displays, topographic maps, and, for some parks, geologic maps prepared by the U.S. Geological Survey. Park interpreters lead excursions, give talks, show slides, and answer questions.

Collecting specimens of any kind is prohibited in national parks and monuments. Leave your rock hammer and collecting sacks at home. An observant eye, a wondering mind, perhaps a pair of binoculars and some sturdy shoes, are all the tools you need.

In this book, geologic terms are defined where first used, as well as in the glossary at the back of the book. In Part 1 they are printed in **bold** as well. In accordance with National Park Service policy, all measurements—distance, height and elevation, and volume—are given in metric units, followed with English units in parentheses.

PART 1.

OUR GEOLOGIC LEGACY

I. Continents That Drift

Beneath your feet, as you stand upon a California shore, the Earth's crust is about 75 kilometers (45 miles) thick, a tough yet fragile skin protecting you from seething depths. Climb to a mountaintop—Mount Rainier or Mount Whitney, say—and it is thicker: over 100 kilometers (over 60 miles). Sail westward over the Pacific and the crust beneath the sea thins to a mere 10 kilometers (6 miles).

Yet this thin skin—and it *is* thin, relative to the 13,000-kilometer (8000-mile) diameter of the Earth—is cohesive and surprisingly flexible. Moved slowly, over vast eons measured in millions of years, it bends and twists. Moved more rapidly, it breaks. Parts of the crust jostle and separate and shift about, sinking or rising, sliding over one another, spreading apart and crunching together, on a scale nearly incomprehensible in size and time.

Fortunately, most of these crustal movements occur at less than a snail's pace, no more than a few centimeters per year. Nevertheless, they affect our lives, for the stresses that cause them may build up

Sudden movement along the San Andreas Fault caused the San Francisco earthquake of 1906. Large parts of the city fell to the trembling earth; other parts were destroyed by fires that raged out of control because the city's water mains were broken.

USGS photo

USGS photo

The double line of the San Andreas Fault runs northward across San Luis Obispo County, California. Movement is largely horizontal, with the west side (left) moving north relative to the east side (right).

over many years, to be released all at once as strong earthquakes or cataclysmic volcanic eruptions.

The rocks of the Earth tell us that our island planet and the continents upon it have suffered similar cataclysms in the past, as well as slower movements that may have little affected the Earth's denizens. In places, rocks are visibly folded, bent, and broken. Elsewhere lava and volcanic ash remain to tell us of past eruptions. We have undeniable proof that during the 4.6 billion year history of our Earth, whole mountain ranges have come and gone, great valleys have filled in or been lifted into mountains, continents have broken and drifted apart. The Earth is ever mobile, heat from within causing constant and dynamic change on its exterior.

By reading the evidence in the rocks, on both local and continental scales, we have learned much—but certainly not all—of the Earth's story. The **crust** is the thin outermost shell of the Earth, the hard part that lies above a thick, incandescent, partly molten and partly solidified layer called the **mantle.** Below the mantle is the core—again partly liquid, partly solid, seemingly composed of iron and nickel. We are of course most concerned with the crust.

Under the oceans, the crust is fairly thin and consists of **basalt.** This black, heavy, iron-rich rock contains many crystals of an olive green mineral called olivine, so the rock is known as **olivine basalt.**

Continental crust is formed of other kinds of rock, most of them lighter in both color and weight: sandstone and shale, granite and limestone, and a whole host of others, many of them common in national park areas. There is hardly any olivine basalt. The continental crust, moreover, is up to ten times as thick as the oceanic crust, one reason why continents rise above ocean basins.

Although they could look at rocks and describe them, or even explain them on local scales, early geologists—right up through the 1950s—were stumped by the problems they encountered in the Earth's crust as a whole. Some suspected, because of similarities in their coastlines, that South America, Africa, and perhaps other continents may have once been joined together. Others noticed a continuity in rock types between parts of America and Europe, or South America and Africa, and took that continuity as evidence of formerly united continents. Many described the long, seemingly pushed-up mountain chains bordering the Pacific

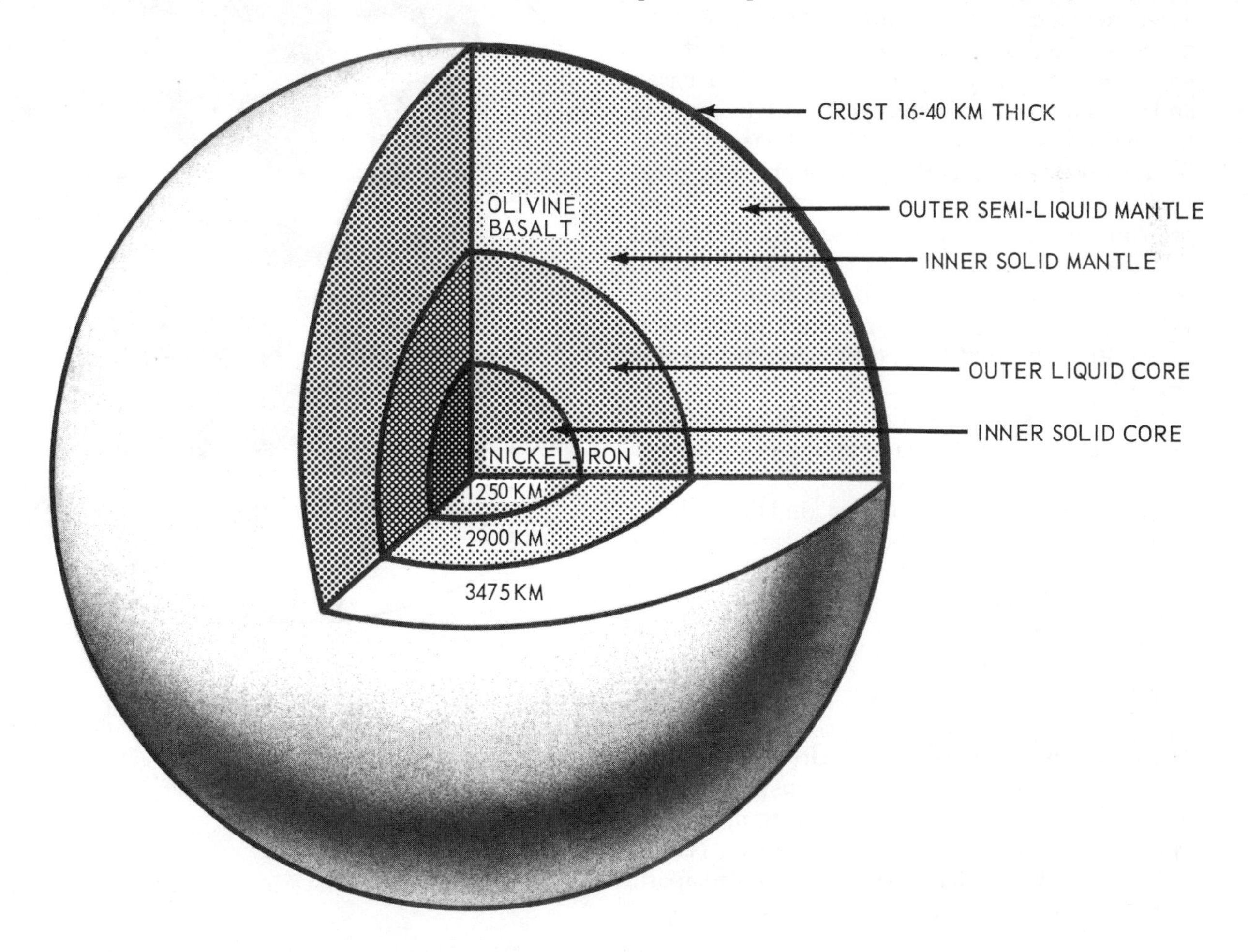

Ocean, as contrasted with the broad, nearly level lowlands that surround the Atlantic. But no one could really explain how parts of the thin, fragile crust could divide and drift around on the semisolid mantle that underlies it.

Then—a geologic revolution! In the light of some brand new ideas, old objections to and repudiation of the concept of continental drift were suddenly swept away. A whole new understanding of the Earth was created almost overnight, and suddenly the answers to myriad geologic questions began to slip into place.

The concept that brought about this revolution is called the **Theory of Plate Tectonics.** According to this theory—now proven beyond any reasonable doubt—continents do indeed drift, but *with* the underlying mantle rather than *across* it. Floating on the hot, semi-fluid upper mantle, the continents are rafted along by great slow-moving convection currents in the upper mantle, currents powered by the superheated stove of the Earth's interior. With a rolling motion like that of boiling soup, huge hot plumes of the mantle slowly rise, roll over, and plunge again into the depths, carrying the continents with them in the upper arches of their rolling motion.

The new theory gives us answers to many of the most puzzling features of the Earth's crust. It tells us, for instance, why oceanic crust is made of basalt and why most continental crust is not, why many of the world's mountain ranges parallel nearby coasts (witness the area covered by this book), how great granite-cored ranges like the Sierra Nevada have come into existence, why we have earthquakes and why they are more frequent in some regions than in others, why there are deep ocean trenches, how volcanoes come into being, and why there is a volcanic "Ring of Fire" (part of it described in this volume) around the Pacific.

The theory states that the crust is made up of individual **plates,** seven large ones and a dozen or more smaller ones. Like the plates of a turtle shell, crustal plates meet along various kinds of sutures. Some sutures are submerged mid-ocean ridges that extend for thousands of kilometers along the floors of all the major oceans. Some are deep-sea trenches like those near Japan, New Zealand, and South America's west coast. Some are great ranges of mountains like the Himalaya. New crust forms along some of the sutures—the mid-ocean ridges—as molten lava wells up from the mantle and hardens into rock. The new crust, created in narrow bands, is then pulled apart by the rolling motion in the mantle, allowing still newer lava to well forth, in turn to harden and to be pulled apart, as if there were two wide conveyor belts moving in

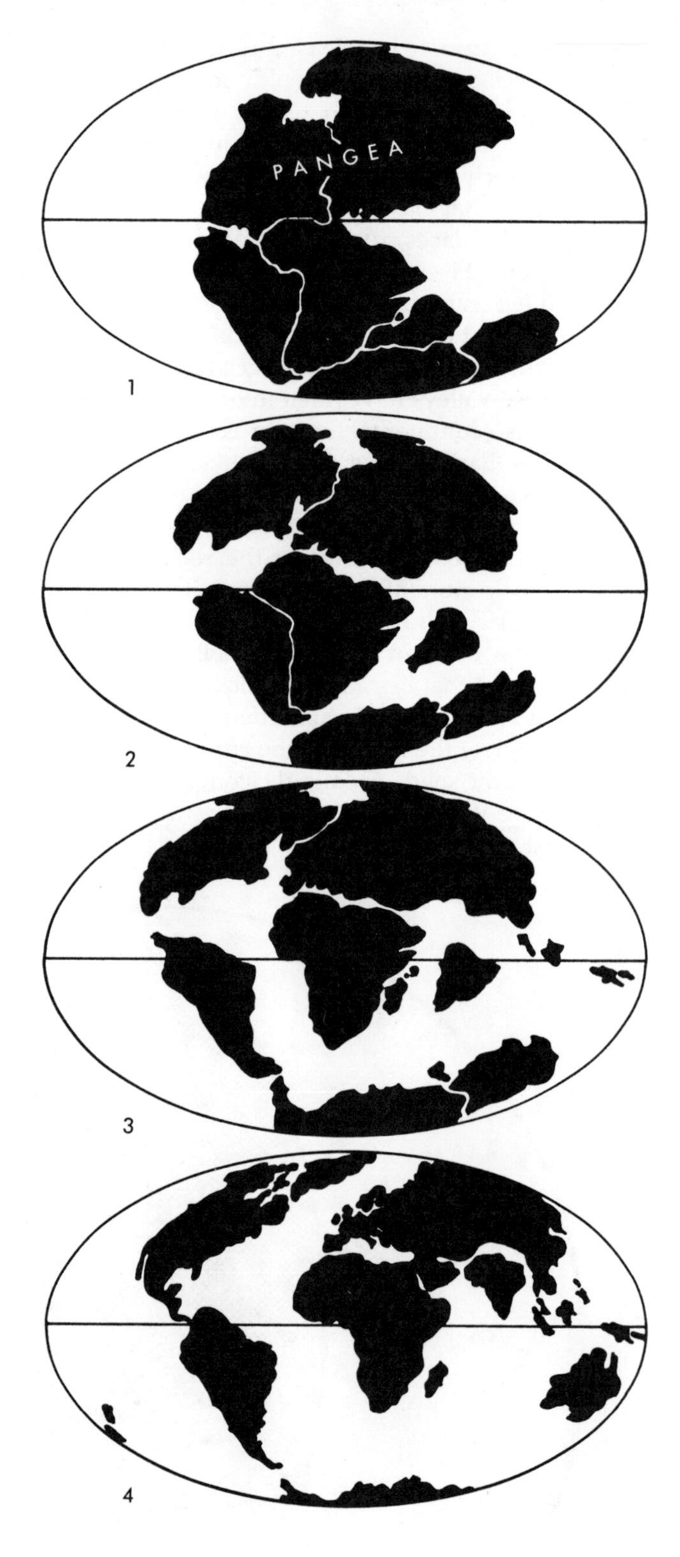

Pangaea, a continent of the past, fragmented to form our present continents. The Americas drifted westward as the Atlantic Basin widened — a movement that profoundly affected the geology of the western United States.

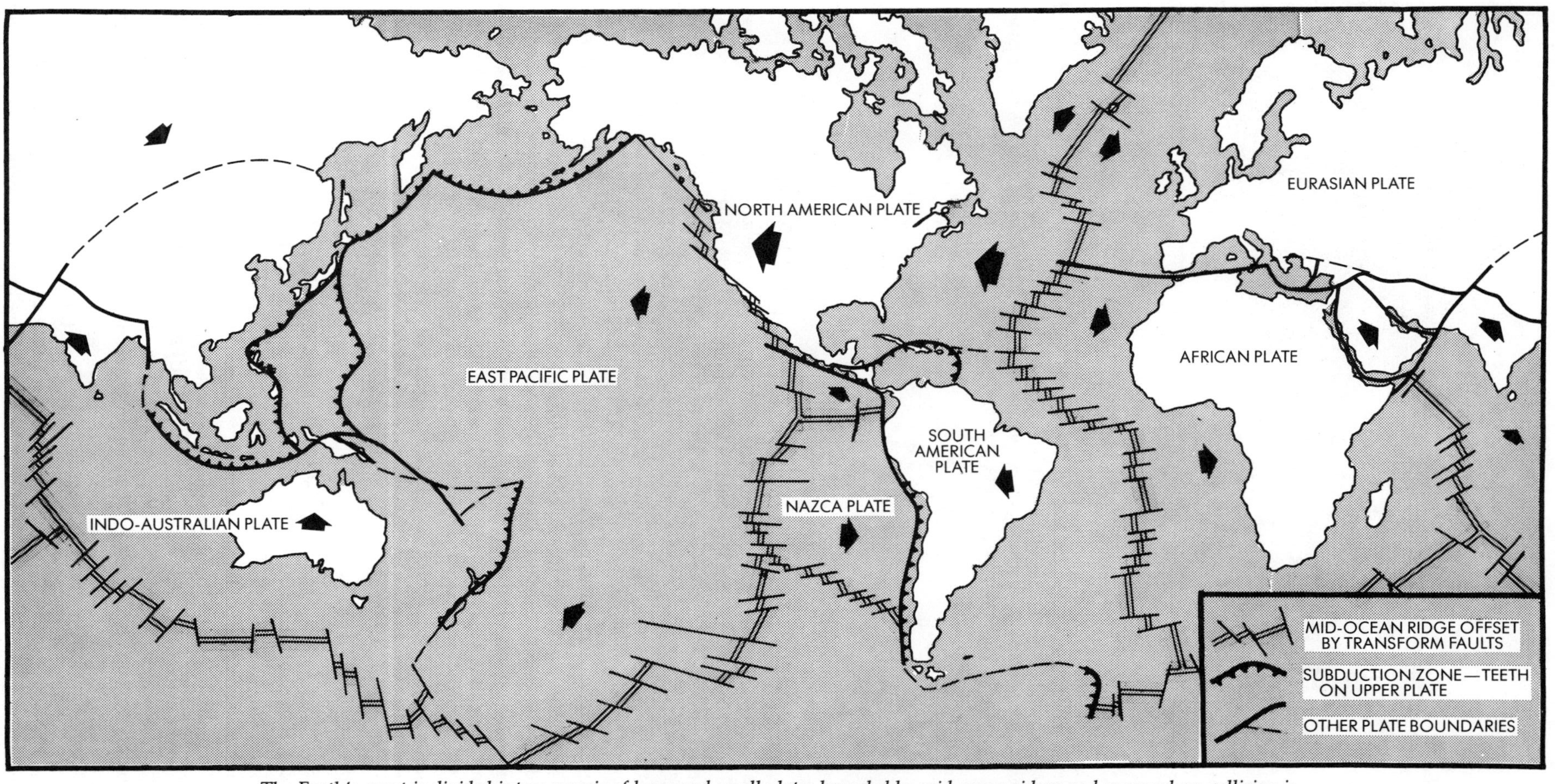

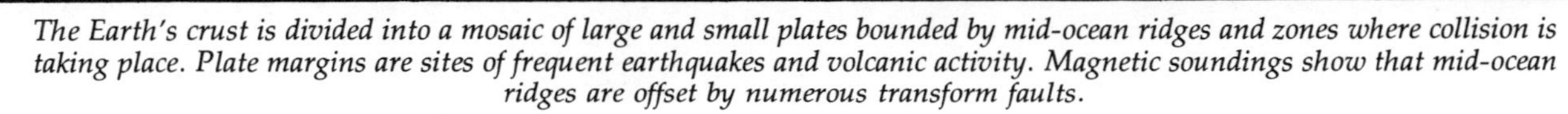

The Earth's crust is divided into a mosaic of large and small plates bounded by mid-ocean ridges and zones where collision is taking place. Plate margins are sites of frequent earthquakes and volcanic activity. Magnetic soundings show that mid-ocean ridges are offset by numerous transform faults.

The design of the Earth's crust is reflected in our national parks. In this diagram, the ocean is narrowed to a fraction of its real relative size.

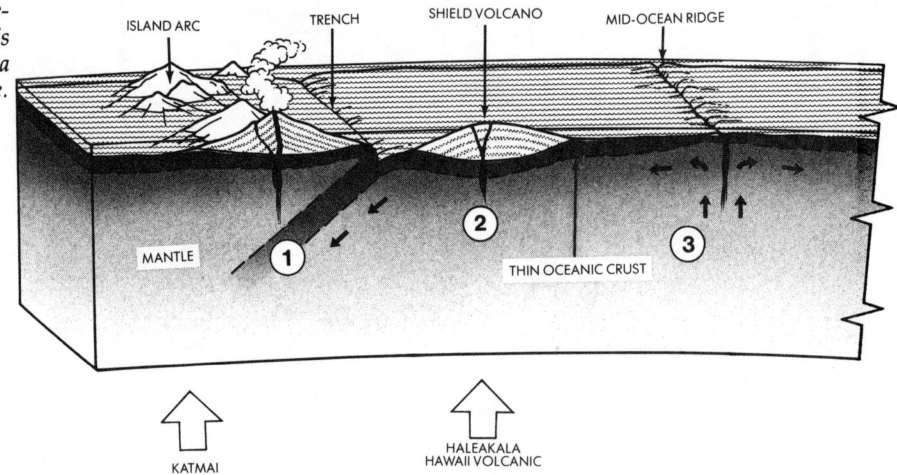

1. *Carried along by convection currents in the mantle, oceanic crust is drawn under and melted in subduction zones, forming basalt magma which erupts along island arcs.*
2. *Shield volcanoes form as plumes of basalt magma rise above isolated "hot spots" below the ocean floor.*
3. *At mid-ocean ridges, basalt magma rises to form new oceanic crust.*

opposite directions. So, as the sea floor spreads, the continents are rafted apart. Their rate of movement varies, averaging between 2 and 5 centimeters (1 to 2 inches) a year.

As new oceanic crust forms at mid-ocean ridges, and as sea floors spread apart, old oceanic crust is pushed toward the continents. There, as the plates collide, the oceanic crust is destroyed.

An oceanic-continental collision has been, as we will see, important to the development of the California, Oregon, and Washington that we know today. The North American Plate, continental in its western portion, has in the last 100 to 150 million years drifted westward with a great convective cell in the mantle. Long ago it collided with the East Pacific Plate. Because its continental rock is lighter and thicker than the heavy basalt of the ocean floor, it has overridden the East Pacific Plate and thereby created much of the scenery described in this book. The North American Plate has now overridden almost all of the East Pacific Plate, right out to the Pacific's mid-ocean ridge, the East Pacific Rise. The tattered edge of the East Pacific Plate shows up as a number of smaller plates, including the Juan de Fuca and Farallon Plates off today's Northwest coast.

The pressures engendered by this overriding have strongly modified the advancing edge of the North American Plate. In places the continent has buckled and broken. Parts of it have been shoved eastward over other parts. Thick layers of mud and volcanic rock, plowed up by the advancing edge of the continent, have in places piled up along its margin. The advancing continent also collected bits and pieces of other continental crust—one-time islands, perhaps—that can be recognized today because they don't match up geologically with the main continental mass. And because the Pacific Basin is also rotating counter-clockwise relative to the North American Plate, the collision has engendered a large fault system, the San Andreas Fault and its many branches, on which movement is mainly horizontal rather than vertical.

At the same time, the **subducted** (drawn down) Pacific Plate, along with slices of continental crust that were pulled under with it, has remelted deep down within the mantle. The melting caused more scenery: Where oceanic and continental crust melted, plumes of molten rock bubbled up again through the edge of the continent to create landscapes of high granite mountains—the Sierra Nevada and Transverse Ranges of California—and the Cascade volcanoes of northern California, Oregon, and Washington.

A similar pattern is seen all around the Pacific. Subduction and melting of oceanic plates have

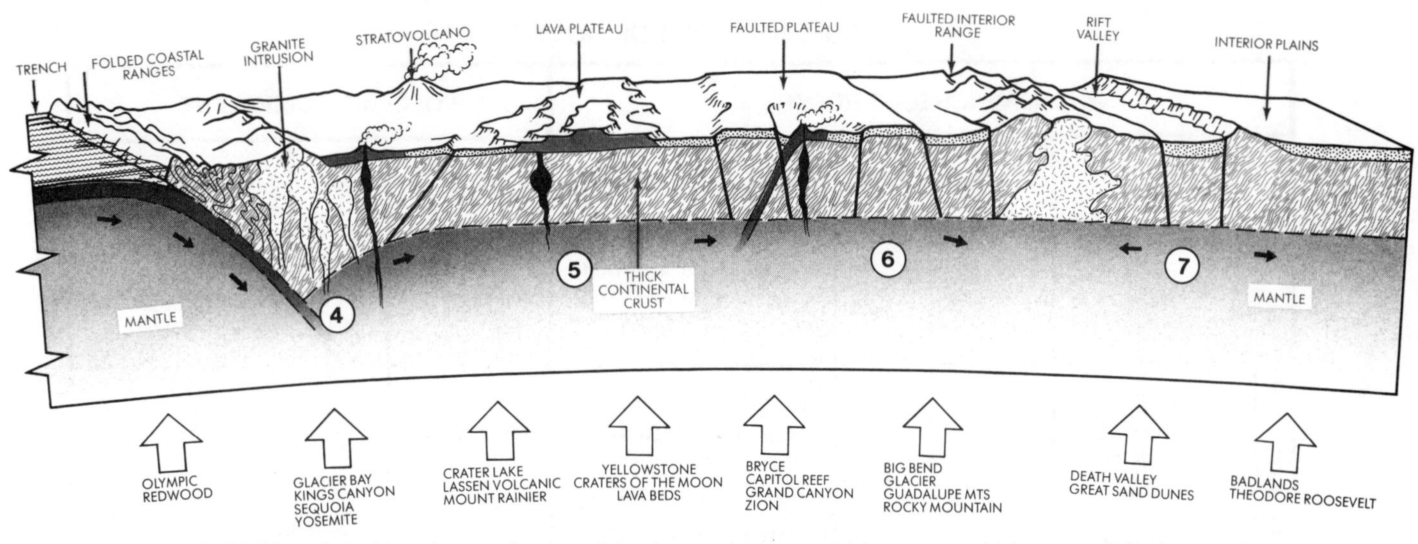

4. *Melting of continental crust along a subduction zone creates granitic magma, which may cool slowly, below the surface, in batholiths that are later bared by erosion. Or the magma may erupt explosively to form stratovolcanoes.*
5. *Flood basalts, exceptionally fluid in nature, rise above "hot spots" below continental crust.*
6. *Interior ranges push upward in response to compression along the distant continental margin.*
7. *Rift valleys are tensional features where the crust drops as neighboring areas are pulled apart.*

given rise to a circlet of volcanoes that stretches from the tip of South America to Alaska, along the Aleutian Islands to Asia, and south again via Japan and the Philippines to New Zealand—the famous (or infamous!) "Ring of Fire." Recent eruptions of Mount St. Helens are tiny blazes along this great ring.

II. Rocks, Time, and Fossils

Rocks make scenery, and even when they are not the main attraction they are important parts of scenery. They and the soil formed from them influence other aspects of scenery: streams, lakes, forests, wildflowers, and animal life.

Geologists recognize three main classes of rocks, all of which occur in the national parks and monuments discussed here:

• **Igneous rocks** originate from molten rock material, **magma,** rising from as much as 300 kilometers (200 miles) below the surface. Igneous rocks are further divided into two main groups: **intrusive igneous rocks** that cool and harden *below* the surface, and **extrusive igneous rocks** that cool and harden *at* the surface. Though the two types may be chemically similar, they look different because slow cooling below the surface leads to the growth of large crystals, so the rocks produced are always grainy or **granitic** (granite-like). Almost all the rocks of the Sierra Nevada fall into this category.

Most extrusive igneous rocks, also called **volcanic rocks,** are fine-textured as the result of rapid cooling, so fine-textured that you can't see individual crystals without magnification. Scattered visible crystals known as **phenocrysts** occur in some volcanic rock, within the fine-textured background. Most of the rocks in the Cascades, with the exception of the North Cascades, are volcanic.

Magma, when it reaches the surface, is called **lava.** When flowing lava pools and cools without further movement, long shrinkage cracks may form at right angles to the cooling surfaces. The cracks meet one another at angles, and so delineate the polygonal (usually six-sided) rock columns of **columnar basalt.** Basalt lavas that erupt under water take on an unusual but readily recognized form, too, looking like stacked pillows and therefore called **pillow lava** or **pillow basalt.**

Another thing to remember about volcanic rocks—and it is admirably expressed in the region discussed in this book—is that differences in the chemistry of volcanic lavas have a profound effect on their tendency to flow or not to flow. Lavas low in silica tend to be much more fluid than lavas high in silica, as we will see when we discuss volcanoes.

COMMON IGNEOUS ROCKS

<table>
<tr><th colspan="5">INTRUSIVE IGNEOUS ROCKS</th><th colspan="3">VOLCANIC EQUIVALENT</th></tr>
<tr><th colspan="2">NAME</th><th>COLOR</th><th>COMP.</th><th>MINERAL MAKE-UP</th><th>NAME</th><th>FLUID.</th><th>OCCURRENCE</th></tr>
<tr><td colspan="2">GRANITE</td><td>light gray to white</td><td rowspan="5">increasingly silicic</td><td>mostly quartz and feldspars, with minor mica and/or hornblende</td><td>RHYOLITE</td><td rowspan="5">increasingly fluid when molten</td><td>fine, pale ash and tuff</td></tr>
<tr><td rowspan="2">GRANITE</td><td>QUARTZ MONZONITE, GRANODIORITE</td><td>light gray</td><td>mostly feldspars and quartz, with mica and hornblende</td><td>RHYODACITE</td><td>light gray ash or tuff, some volcanic domes</td></tr>
<tr><td>QUARTZ DIORITE</td><td>medium gray</td><td>mostly feldspars, with quartz, mica, and hornblende</td><td>DACITE</td><td>gray lava of volcanic domes, ash or tuff, pumice</td></tr>
<tr><td colspan="2">DIORITE</td><td>medium gray</td><td>feldspars and quartz with abundant dark minerals</td><td>ANDESITE</td><td>short, thick lava flows, volcanic ash or tuff</td></tr>
<tr><td colspan="2">DIABASE</td><td>dark gray to black</td><td>mostly dark minerals such as pyroxene, some plagioclase and quartz</td><td>BASALT</td><td>fluid black and dark gray lava flows, cinders</td></tr>
</table>

COMMON SEDIMENTARY ROCKS

NAME	DESCRIPTION
MUDSTONE	grains of silt and clay cemented together
SILTSTONE	grains of silt cemented together
SHALE	siltstone or mudstone that splits into flat slabs parallel to original bedding
SANDSTONE	grains of sand (usually quartz) cemented together
CONGLOMERATE	sand, pebbles, and cobbles deposited as gravel and later cemented together
LIMESTONE	calcium carbonate (calcite) rock deposited as limy mud or fragments of shells

COMMON METAMORPHIC ROCKS

NAME	DESCRIPTION	DEGREE OF ALTERATION
SLATE	altered mudstone or siltstone that fractures along planes not parallel to stratification	slight
QUARTZITE	sandstone or siltstone so tightly cemented that it breaks through individual grains	slight
MARBLE	recrystallized limestone or dolomite	slight to severe
GNEISS	banded or streaked crystalline rock formed by recrystallization of older granite, sandstone, or sandstone-shale layers	severe
SCHIST	medium-grained rock formed by alteration of mudstone and siltstone, with aligned mica grains that cause it to split along parallel planes	severe
GREENSTONE	an old term applied to hard, green metamorphic rock derived from dark igneous (usually volcanic) rocks	severe

Low-silica lavas are generally much darker in color than high-silica lavas. (The same can be said of intrusive igneous rocks.)

• **Sedimentary rocks** form from broken or dissolved pieces of other rocks. They are layered, or **stratified,** and so collectively they are called **strata.** Some volcanic rocks—lava flows and widespread sheets of volcanic ash—are stratified too, especially when they are interlayered with sedimentary rocks.

Most sedimentary rocks are given names which reflect the grain size of their original components, as shown on the accompanying table. Limestone is an exception; it forms from the calcareous shells of shellfish, or from **calcium carbonate** ($CaCO_3$) chemically precipitated from sea water.

• **Metamorphic rocks** come into existence when pre-existing rocks are altered by pressure and/or heat. In those formed from sedimentary rocks, individual sand or mud grains may fuse together, as is the case with **metasedimentary rocks,** which look like much harder versions of the sandstones and mudstones of which they were made. Or they may be altered so much that the original materials have recrystallized into new minerals, as happens with **marble.** Much-altered metamorphic rocks may look quite a bit like intrusive igneous rock, so it is frequently difficult to draw the line between them. Since intrusive rocks and highly altered metamorphic rocks both show visible mineral crystals, it is sometimes handy to get around this problem by speaking of them collectively as **crystalline rocks.**

Rocks are made of **minerals,** natural substances with definite chemical make-ups. Minerals very often come in recognizable and characteristic colors, and have typical hardnesses and characteristic ways of crystallizing. For instance, the mineral **quartz,** colorless and glassy when pure, has the chemical composition SiO_2: one atom of silicon attached to two atoms of oxygen. Quartz is hard enough to scratch glass or steel, and where it has room to grow without bumping into other crystals it develops beautiful six-sided crystals with beveled, pointed ends.

There are literally hundreds of different minerals, some much more common than others. A few you probably know quite well already: quartz, mica, native gold, gemstone minerals like ruby and diamond and garnet, and a mineral called water in its liquid phase and ice or snow in its solid or crystal phase. **Halite,** a crystalline white mineral, is recognized by its taste—salt. Minerals are not stressed in this book; if you like to identify them you probably already own one of the guidebooks used by rockhounds. Remember: no collecting in the parks and monuments.

GEOLOGIC TIME

ERA	PERIOD	EPOCH	AGE IN YEARS
CENOZOIC Age of Mammals	QUATERNARY Q	HOLOCENE Q	10,000
		PLEISTOCENE Q	2 million
	TERTIARY T	PLIOCENE Tp	5 million
		MIOCENE Tm	24 million
		OLIGOCENE To	37–38 million
		EOCENE Te	55–57 million
		PALEOCENE Tp	63–66 million
MESOZOIC Age of Reptiles	CRETACEOUS	K	138–144 million
	JURASSIC	J	205–208 million
	TRIASSIC	T̄R	240–245 million
PALEOZOIC Age of Fishes	PERMIAN	Pm	286–290 million
	PENNSYLVANIAN	P or IP	320–330 million
	MISSISSIPPIAN	M	360–365 million
	DEVONIAN	D	408–410 million
	SILURIAN	S	435–438 million
	ORDOVICIAN	O	500–505 million
	CAMBRIAN	Ꞓ	570 million
PRECAMBRIAN P-Ꞓ	ORIGIN OF LIFE		2.5 billion
	ORIGIN OF EARTH		4.6 billion

That brings us to time. For geologists, time goes back 4.6 billion years, to the creation of the Earth, or 3.8 billion years, when the oldest rocks now known, metamorphic rocks of central and eastern Canada, were formed.

Geologists only very recently learned how to measure geologic time with a fair degree of accuracy, something that can be done by analyzing the breakdown of radioactive minerals in rocks, or by relating natural rock magnetism to the now-known history of reversals in the Earth's magnetic field, when the North Pole and South Pole switched their positive and negative magnetic charges. Before these methods were discovered, geologists invented their own calendar, with special names for the months, weeks, and days of geologic time, names that are still used because to geologists they are the days of the week and the months of the year.

The largest units, the geologic months, are called **eras.** Eras are divided into **periods,** just as months are divided into weeks; periods are split up into **epochs,** which we can liken to days. Because Nature is less precise than Man, and because great chunks of the record of the past are missing completely, eras and periods and epochs are not as regular in length as months and weeks and days. The longest era lasted about 4 billion years, the shortest only 65 million.

Names for the time units are shown on the opposite page, along with abbreviations commonly used on maps and diagrams. Approximate ages in years are given as well. The dates shown here differ slightly from those in Volume 1 of *Pages of Stone,* reflecting recent research and refinement of dating techniques. Names of eras, the major divisions, are derived from Greek words for "old"—**paleo,** "middle"—**meso,** and "new" or "recent"—**ceno.** The **zoic** parts of the names refer to "life" (as in zoo).

Paleozoic periods are named from places where rocks of particular periods were first studied. Cambria was the Roman name for Wales, and the Ordovices and Silures were two tribes encountered there by Roman armies, way back in 74 to 78 A.D. Devon is the southwestern peninsula of England that lies just south of Wales. The next two periods are American: Mississippian and Pennsylvanian. (Europeans call the same time interval Carbonifer-

Poorly consolidated gravel of the Gold Bluffs Formation forms sea cliffs above a beach in Redwood National Park.

ous because rocks of that age contain lots of coal, both in Europe and in eastern United States.) Permian comes from Perm, a province in the Ural Mountains of Russia.

As for the Mesozoic periods, Triassic is called Triassic because in Germany rocks of this age are made up of three distinct layers. Jurassic rocks occur in the Jura Mountains of Switzerland. Cretaceous means "chalky" and refers to the thick chalk layers of the coasts of England and France.

Cenozoic period names trace their ancestry back to some old and now-disproven ideas that divided rocks into four groups by their relative degree of consolidation. The groups called "Primary" and "Secondary" fell by the wayside, but Tertiary, for poorly consolidated rocks, and Quaternary, for distinctly unconsolidated rocks, stayed in the geologic vocabulary. With sedimentary rocks, it's true that the youngest layers are usually poorly put together sand, clay, and gravel, whereas the older ones become compressed and cemented with age. We know now that hardness does not necessarily connote age: Many igneous rocks are hard the minute they cool.

Tertiary and Quaternary rocks are near the surface and haven't had much opportunity to get squeezed, heated, eroded away, or covered up by other rocks, so we know more about them than about older layers. Because of this, Cenozoic epoch names are useful, though they are used sparingly in this book.

Most of the rocks described in this volume are Mesozoic or Cenozoic—there just aren't many Paleozoic or Precambrian rocks at the surface in the far western states. They are down there underneath, however, at least where they were not eroded away or subducted and melted beneath the west-drifting continent.

Regardless of their age or origin, rocks as they occur in the landscape divide into recognizable natural units which geologists call **formations.** These units are distinguished by their composition, thickness, color, internal structure, mineral or fossil content, and appearance in outcrop. They are named after a locality at which they occur, called a **type locality,** which then becomes the standard for comparing other parts of the same formation.

Nature has not been kind to classifiers. Formations, which are *rock* units, may transgress *time* units such as periods and epochs. A formation made of beach sand, for instance, advances in time as well as place as the sea encroaches farther and

Metamorphic rocks created from older sedimentary and volcanic layers reflect the high degree of metamorphism that created the Skagit Gneiss, in North Cascades National Park.

farther across a sinking shore. Formations also may grade in composition from place to place, being, for example, pure marine limestone in one area but interlayered with silty shore deposits in another. Similarly, they may change in composition from bottom to top as their environment changes with time, as with a decrease in sand or mud provided by rivers or an increase in stream pebbles and cobbles reflecting uplift of nearby mountains.

We think now that life came into existence about 2.5 billion years ago, as simple, minuscule noncellular and one-celled organisms living in seas or lakes. Only gradually did these tiny specks of living material evolve into more complicated life forms in which many cells were clustered together as primitive plants and animals.

About 600 million years ago, some of the more complex multicellular varieties, as well as some single-celled forms, began to make themselves protective shells. Some secreted for this purpose a substance called **chitin,** horny material like that in crab and lobster shells. Others learned to secrete calcium carbonate, the rigid rock-like white stuff of most seashells. Yet others secreted **silica,** forming exquisite and delicate glassy shells. Shells gave protection from predators and scavengers, and also, with the demise of their owners, granted a special favor to geologists: Through eons of time, shells collected on sea bottoms and in other places became buried by sand, mud, and ooze, the material that would consolidate into rock. Found much later, they came to be called **fossils** (from the Latin *fodere,* to dig).

There are not very many fossils in Precambrian rocks, partly because those rocks have had such long histories of folding and breaking and metamorphism, and partly because living things had not yet begun to develop shells. But we are sure there *were* living organisms, at least in the last part of Precambrian time. The complex Cambrian fossil fauna and flora, representing most of today's major groups, could not have sprung into existence full-blown. They must have descended from shell-less Precambrian ancestors.

During the Paleozoic Era, sometimes called the "Age of Fishes," there were shellfish of many kinds, including relatives of corals, clams and snails, crabs and lobsters. After the middle of the era, there were increasing numbers of fishes, the first animals to have jointed backbones. There were plants, too, and after the middle of the era some of the plants and then some of the animals left the sea and took up life partly or wholly on land.

The Mesozoic Era saw rapid evolution of land-living animals, particularly the newly evolved reptiles, so it is called the "Age of Reptiles." Turtles, crocodiles, swimming reptiles called **ichthyosaurs,** flying reptiles known as **pterosaurs,** and the giant dinosaurs swam, flew, and roamed around the world. During this era two other groups of vertebrate animals appeared: birds and mammals.

The dinosaurs, pterosaurs, and ichthyosaurs vanished quite suddenly at the end of the Mesozoic Era, along with about half of the animal and plant species that then existed—a mass extinction thought to have been brought about by an asteroid or comet collision that so filled the atmosphere with dust that the Earth's surface was plunged into darkness for many months. As life recovered from this extinction, the mammals, no longer threatened by the once-abundant giant reptiles, rapidly diversified and spread over the land, into the air, and into the sea to fill the just-vacated ecologic niches.

Why the mammals should survive while so many reptiles died out we do not yet know; perhaps their fur-covered bodies, high body temperatures, and ability to suckle their young played a part in their survival. Perhaps they just dug deeper burrows, or were adapted to a nocturnal existence that helped them survive the long, dark night. Anyway, the Cenozoic Era became the "Age of Mammals." Mammals multiplied and diversified through millions of years. And somewhere along the line, just a few million years ago, a group of mammals evolved that could balance on their hind legs, freeing their forepaws to grasp and grab. One particular group also developed large, very skillful, extraordinarily inquisitive brains.

The development of man came about at a time when the Earth was very much as it is today. The continents were approximately where they are now in terms of their distance apart and their positions relative to the Earth's poles and equator. All the major seas and major mountain ranges of today were already in place, though some of the ranges may not have been as high as they are today. Climates were variable—sometimes glacially cold, sometimes warmer than at present, fluctuating in ways well expressed in many of the national parks discussed in this volume.

As mountains are built and glaciers grind them down, as rivers wash them away and seas flood what was once dry land, as volcanoes spurt lava and ash, parts of the Earth's rock-recorded history have been lost—inevitably and irrevocably. From what remains, fragmented though it is, we tell the Earth's story.

III. The Making of Mountains

The mountains and plateaus of Washington, Oregon, and California illustrate five different types of mountain-building:

• **Folded and faulted mountain ranges,** made of very old rocks, like the North Cascades.

• **Intricately folded mountains** made of fairly young sea-floor sediments and lava flows, like the Olympic Range.

• **Plateaus of flat-lying lava flows** like the Columbia River and Modoc Plateaus.

• **Individual volcanoes** like Mount Rainier, Mount St. Helens, and Lassen Peak.

• The great **fault-block range** of the Sierra Nevada.

As in the rest of the world, all these mountain structures can be laid, directly or indirectly, at the door of plate tectonics. We've seen already that the North American Plate drifted west (a process still going on) as the Atlantic Ocean widened, overriding the Pacific Plate. All along the line of the great collision, the edge of the continent was bent and torn and in places pulled down with the oceanic plate into depths where remelting could take place. The bending and tearing and remelting are responsible for the present mountain ranges, intermontane basins, and volcanoes.

In some of the ranges thus created, horizontal pressure dominated, and the crust was squeezed together and crumpled until it occupied less east-west space than it did before. Elsewhere, horizontal tension pulled the crust apart into separate blocks, which tipped and tilted like so many jumbled dominoes. Where the East Pacific Plate was subducted as the continent slid over it, sea-floor basalt and mud were scraped off the descending oceanic plate as if by a giant bulldozer, and piled on the western edge of the continental plate, the bulldozer's blade. Along parts of its margin, the west-drifting continent added to itself some oceanic islands—scraps of rock unrelated to the rest of the continent. Elsewhere magma—molten rock made of dragged-down, deep-melting sea-floor and continental material—collected in huge incandescent masses in magma chambers below the edge of the continent, cooling there or working its way to the surface to erupt as volcanoes.

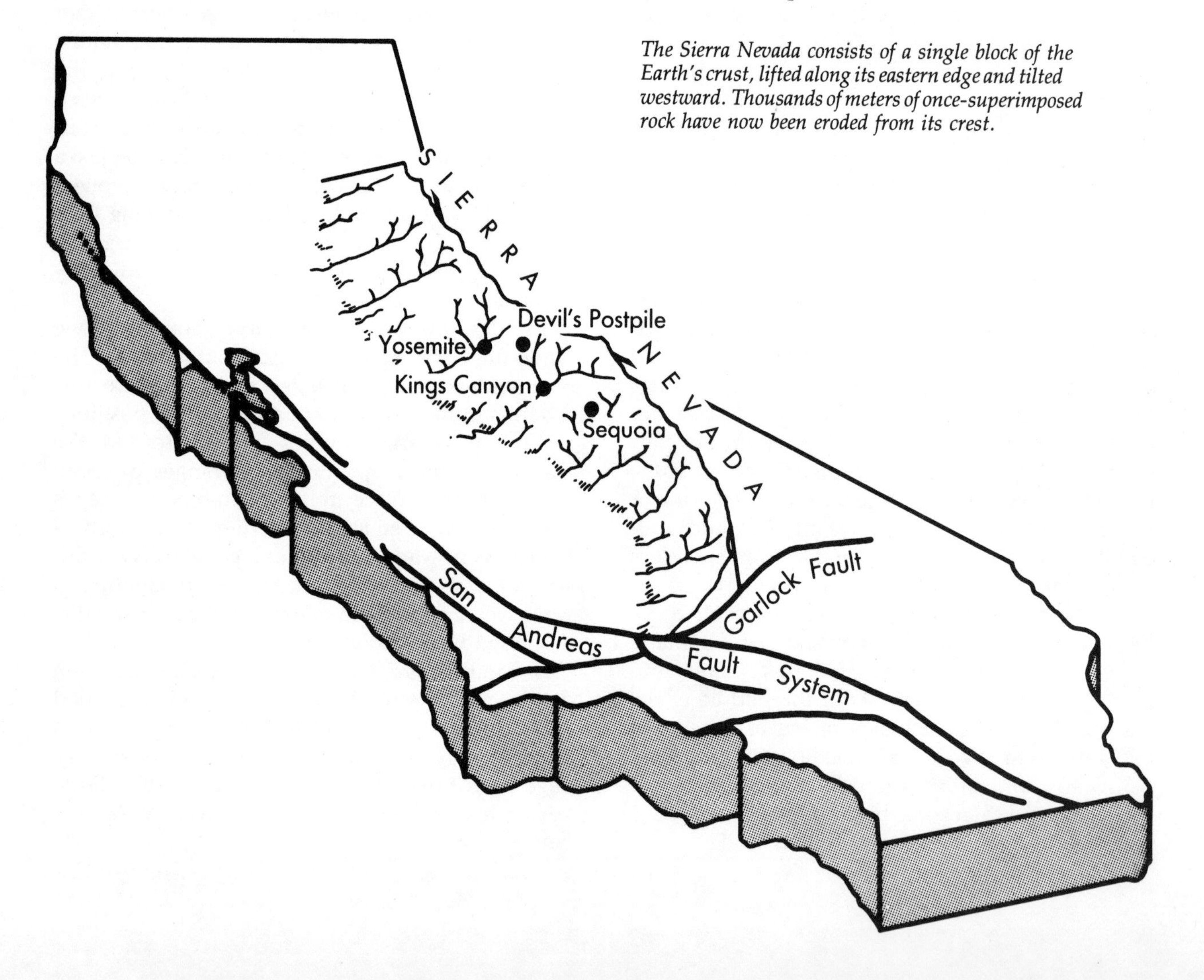

The Sierra Nevada consists of a single block of the Earth's crust, lifted along its eastern edge and tilted westward. Thousands of meters of once-superimposed rock have now been eroded from its crest.

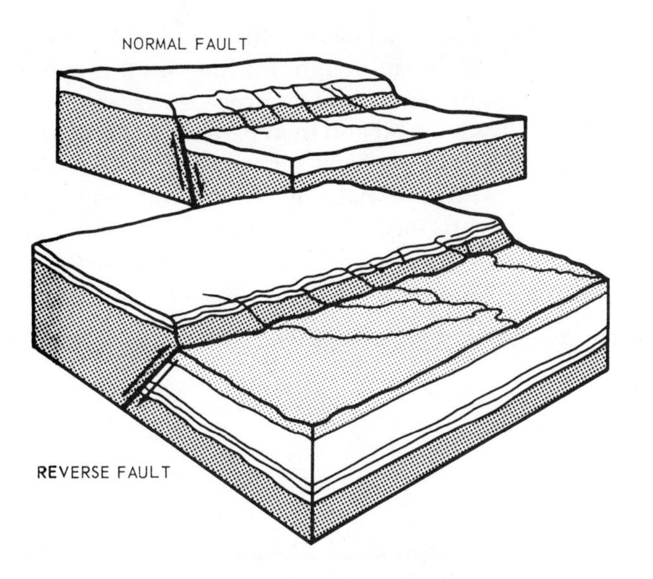

Geologists have found it convenient to have precise terms to describe features caused by the bending, folding, and breaking of rocks. An upward arch, on any scale, is an **anticline;** a downward bend is a **syncline,** as illustrated here. In eroded anticlines, the oldest rocks are near the center; in eroded synclines, the youngest are near the center—important relationships for those doing geologic mapping. Both anticlines and synclines may be tightly compressed, accordion pleated. Anticlines in which rocks slope away in all directions from a center are called **domes;** large equidimensional synclines are **basins.**

Nearly all exposed rocks show signs of breakage, either in the form of cracks or **joints** along which there is no relative movement of the two sides, or in the form of **faults** along which there *is* relative movement. High-angle (nearly vertical) faults may be **normal faults,** with the upraised block not overhanging the downthrown block, or **reverse faults,** with the upraised block overhanging the

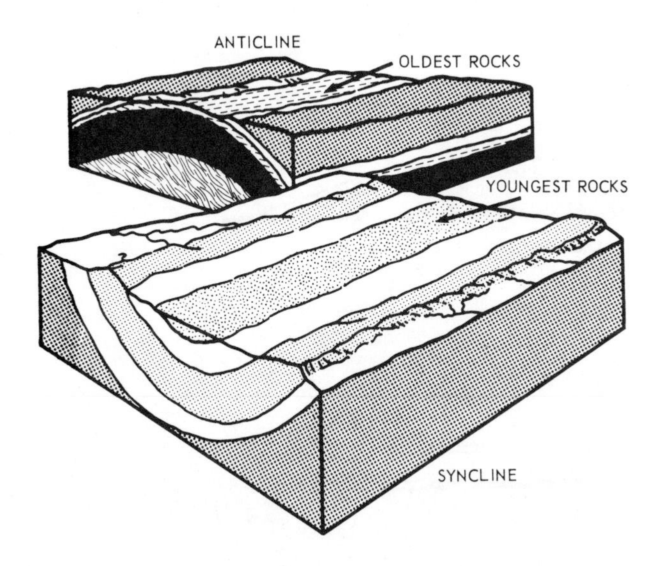

downdropped block. (In real life, of course, erosion does away with the overhang.) In low-angle (nearly horizontal) faults, usually called **thrust faults** or **overthrusts,** the upper block is thrust across the lower block, as the name implies.

Faults of all kinds commonly occur not as single clean breaks, but as zones of disrupted rock. Because such **fault zones** are lines of weakness, they may show up as long, straight valleys, lines of springs, or lines of volcanoes.

It may seem strange that rock—solid rock—can bend and break. The necessary factors seem to be stress, pressure, temperature, and time, and the more of these factors the better. Stress comes, as we've seen, with the continent's westward drift and collision with the East Pacific Plate. It may also come from below, because of the rise of magma. Pressure comes from deep burial under thousands of meters of overlying rock. Temperature, too, comes with depth and with rising magma. And there is, as we have seen, plenty of time.

Both volcanic plateaus and the cone- or dome-shaped hills and mountains commonly recognized as volcanoes develop around volcanic **vents** where molten magma reaches the surface through fissures (joints or faults) or isolated, narrow, pipe-like conduits. Several types of volcanoes occur in the western parks:

• **Basalt plateaus** occur where very fluid lava flows quietly out over surrounding land, filling valleys and leveling topography, as it has done around Lava Beds National Monument and the high plateaus on which are superimposed Mount Rainier and other Cascade volcanoes. Such plateau-forming lavas are usually basalt, but may have altered with time to **greenstone,** a metamorphic rock.

• **Shield volcanoes** develop when slightly less fluid lava erupts from a central vent. These volcanoes range in size from small hills to the immense broad-based island mountains of Hawaii. There are several small shield volcanoes in Lassen Volcanic National Park.

• **Cinder cones** and **spatter cones** develop if the gas content of lava is very high. Bubbles rise to the top like froth on a glass of beer, splashing bubbly clots of lava around the vent. With greater force, lava fragments shoot higher and fall to earth in a rain of cinder. Some of these small volcanoes occur in Lava Beds National Monument and in Lassen Volcanic National Park.

• **Stratovolcanoes** or composite volcanoes develop when moderately gummy or viscous lava, with composition ranging from andesite to rhyolite, erupts repeatedly in violent bursts that shoot out fine volcanic ash, blobs of molten rock, and chunks of solid rock, with other, quieter eruptions in which lava domes build up in the crater or more fluid lava cascades down the volcano's flanks. Alternate layers of lava and volcanic ash eventually pile up into well defined and often very beautiful cone-shaped, crater-topped mountains. Mount Rainier, Mount St. Helens, and other large Cascade volcanoes come into this category.

• **Lava domes** involve very stiff lava of the consistency of thick bread dough. Pushing up through a volcanic vent, such lava scarcely flows at all; it bulges outward a bit and then hardens in place as a rounded dome. Volcanic domes are usually associated with pre-existing stratovolcanoes, and they are nearly always surrounded by aprons of

Volcanoes of several types are represented in the Cascade area. Volcano shape is determined by the fluidity of lava and the amount of gas it contains.

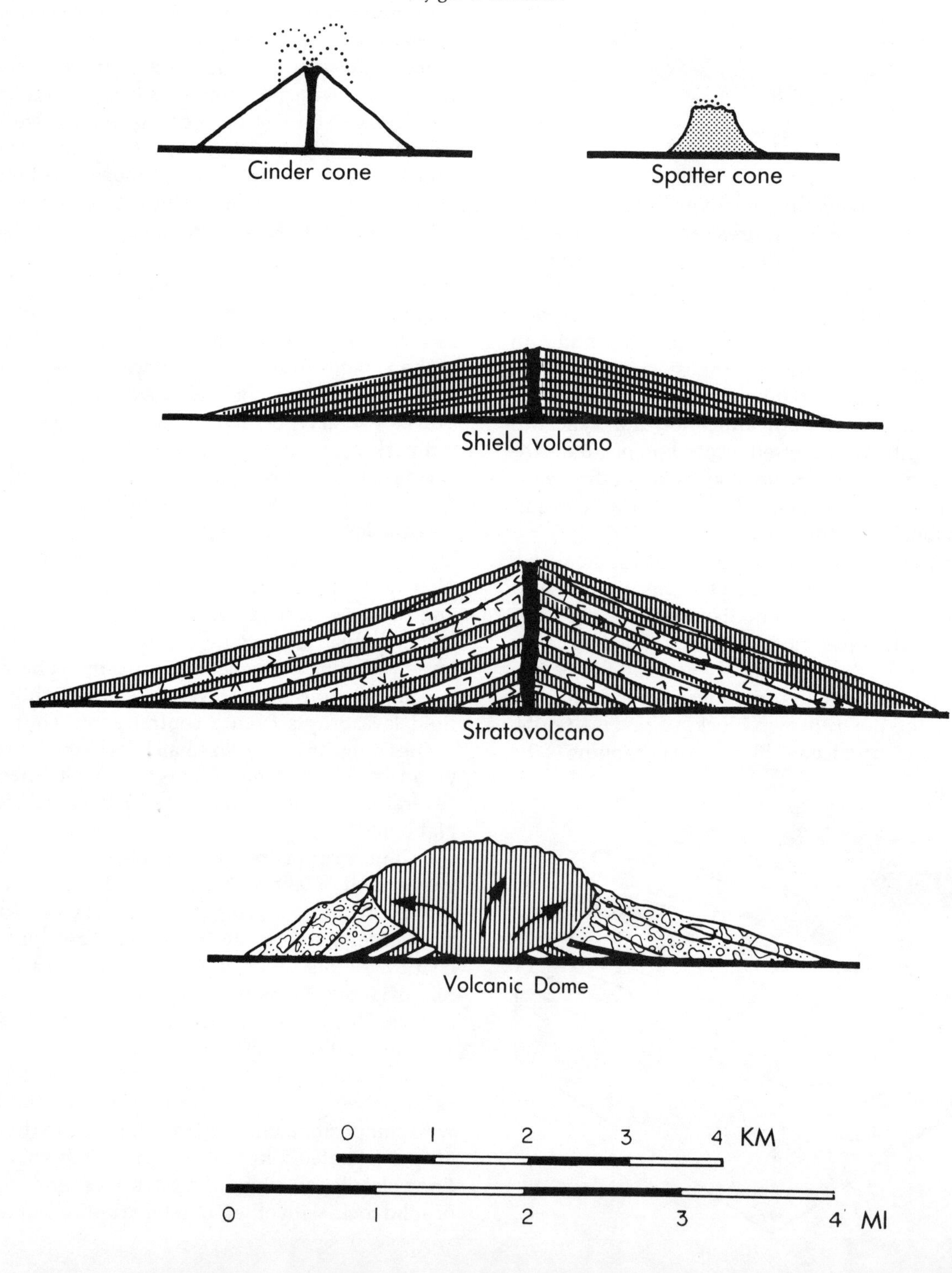

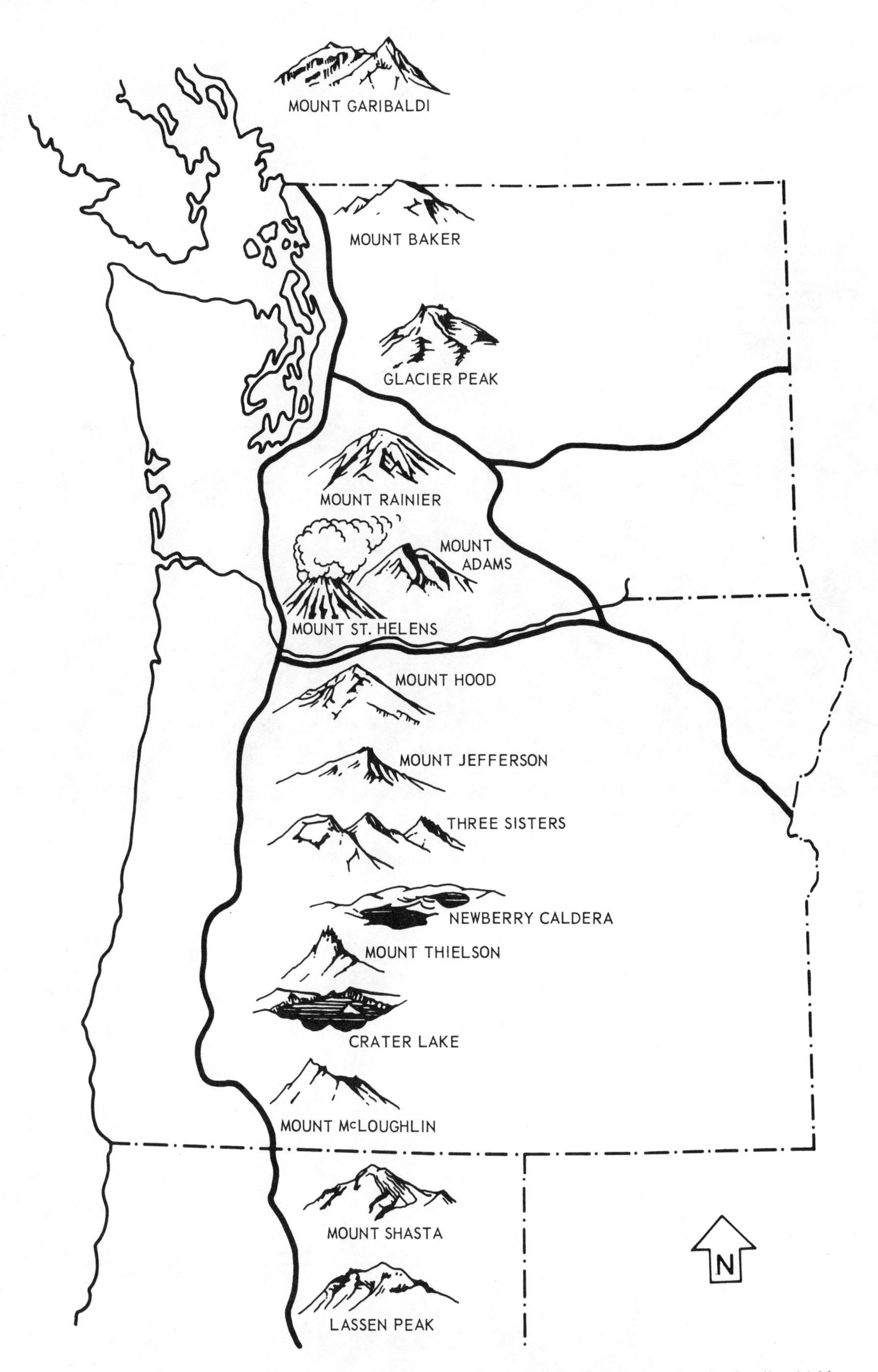

The Cascade volcanoes stretch in a long line from Canada to California. Three are national parks. Smaller shield volcanoes and cinder cones are not shown.

USGS photo, courtesy David A. Johnston Cascade Volcano Observatory

Since Mount St. Helens' great eruption in May of 1980, several steaming lava domes have pushed upward in the volcano's yawning crater. Early ones were blown apart by subsequent eruptions. This dome, a compound of many thick, lobe-like flows, was photographed in January of 1984.

Wizard Island, a volcano within a volcano, rises above the still, deep waters of Crater Lake. The distant caldera rim was part of Mount Mazama.

talus made up of bits and pieces of their own hardening crust loosened by continued swelling of the dome mass. Domes may eventually blow apart because of renewed volcanic activity, as has happened recently at Mount St. Helens, or they may remain in place indefinitely, as several have done at Lassen Volcanic National Park.

Because their lava is thick and viscous, yet gaseous, stratovolcanoes are more lethal than other types of volcanoes. In historic time they have caused catastrophes around the world: the destruction of Minoan civilization in about 1470 B.C., of Pompeii and Herculaneum in 79 A.D., of Indonesian islands near Krakatau in 1883, of the city of St. Pierre on Martinique in 1902. Ash falls from the Cascade volcanoes have many times blanketed the whole Northwest. Mount St. Helens' 1980 eruptions are child's play by comparison with some earlier eruptions, as these figures show:

Date	Volcanic site	Magma volume
May 1980	Mount St. Helens	1 km^3 (1/4 mi^3)
1883	Krakatau	70 km^3 (17 mi^3)
6800 yrs ago	Mount Mazama	50 km^3 (13 mi^3)
600,000 yrs ago	Yellowstone Caldera	1000 km^3 (238 mi^3)

Stratovolcanoes are quite capable of their own destruction, as well. When they blow off enormous quantities of foaming, explosive, gas-rich magma, the level of magma in the magma chambers lowers drastically and the volcanoes not uncommonly collapse, leaving a circular depression called a **caldera.** Mount Mazama, in what is now southern Oregon, 6800 years ago emitted so much ash that it eventually collapsed into its own magma chamber, creating the cliff-walled basin now occupied by Crater Lake. Other signs of a volcano's passing may remain as well—thick beds of volcanic ash, steaming fumaroles from which volcanic gases still rise, hot springs, and geysers. Most of the volcanoes in the Cascade region of Washington, Oregon, and northern California are dormant, sleeping, and as we learned in 1980, may awaken at any time.

IV. The Wearing Away

Sedimentary rocks chronicle repeated periods of deposition when, on sea floors and lake floors and in continental basins, gravel and sand, silt, clay, and calcium carbonate were deposited in horizontal layers. These rocks also chronicle uplift and mountain building and tell of the ceaseless cutting, carving, wearing, carrying, and ultimate depositing that bring rock material from highland to lowland.

Uplift invites erosion. Wherever land rises, it begins to wear away. Streams gushing down steep slopes flow faster than sedate rivers of the valleys, and the faster water is moving, the greater the burden it can carry.

The first step in the erosion process is **weathering**, physical and chemical disintegration of rock. Water, ice, air, temperature changes, and plant and animal matter all play parts, in their special ways, to turn rock into soil.

Water in itself is not particularly corrosive. However rainfall and snowmelt, having absorbed carbon dioxide from the atmosphere, are weakly acid. The dilute acid attacks particularly the calcium carbonate of limestone, a substance that also cements mineral grains of many other sedimentary rocks, and the micas and feldspars of igneous rock. As minerals decompose, the rocks weaken and break down, sometimes grain by grain, into soil—a slow but relentless process. The presence of lichens and other types of plants and plant products speeds up the process. Air moist with mild acids attacks rock in the same way. Acid-enriched water acts below the surface as well, etching and dissolving caverns in limestone and related rocks.

At high altitudes and high latitudes, water works also in its solid form, as ice or frost. Unlike other minerals, water expands as it crystallizes, exerting a powerful pressure against its surroundings. Repeated growth and melting of ice crystals where water has seeped into rock pore spaces dislodges individual grains; development of frost in joints and other crevices wedges apart whole blocks of rock.

Temperature changes on exposed rock surfaces cause expansion and contraction of the rock itself. Especially in the outermost few inches, such expansion and contraction loosen individual mineral grains or thin flakes of rock. Mere unloading of long-buried rock as overlying material is washed away, allows the rock to expand, a process that is accentuated in intrusive rocks that solidified at great depths under immense pressures. Such expansion creates **pressure release joints** concentric to the rock surface, separating large curving slabs of rock and allowing moisture to seep in. Coupled with frost action, both of these processes round boulders and create bare rock domes. Plant roots may grow into these and other joints, also forcing the rock apart, though whether the plant growth itself or the swaying in the wind exerts the necessary force is often hard to determine.

Once loosened or turned into soil, rock material is vulnerable to another process—transportation. The

Weathering along joints may exceed that on unjointed rock surfaces, resulting in grooves such as these. Free-standing, somewhat angular boulders—erratics—were dropped by a melting glacier.

very same erosional processes we see around us today operated thousands, millions and, with the exception of those associated with plant and animal activity, even billions of years ago. They operate on every scale, from transport of molecules in solution in water to transport of great boulders by flooded streams. Again, we are looking at processes involving water, ice, air, and temperature changes, as well as at two other forces: wind and gravity.

Through the ages, rivulets, streams, rivers, and the water of seas and lakes have picked up where weathering left off, transporting broken-down rock material toward its ultimate resting place, the sea, in ways that are familiar to us all. Heavy rains, floods, coastal and mountain storms are potent agents in these processes; they move many millions of tons of rock material each year. But quieter streams and more leisurely rivers, acting through thousands of years, also move the products of weathering, carrying minerals in solution, patiently rolling grains of sand, repeatedly undermining cobbles and boulders until they roll a half turn or a quarter turn downstream. The heaviest floods are mudflows caused by sudden overdoses of water and rock material—commonly in connection with volcanic eruptions. In them, water is so thick with volcanic ash and finely broken rock that it can almost float large boulders and cobbles in its cementlike mass.

Plant life lessens the effectiveness of erosion. The plants themselves shield the ground by intercepting falling raindrops. Their roots hold soil together; their branches collect dust and sand. And when they die, their decomposing stems, trunks, and foliage mix with decomposed rock to form spongy, water-absorbing soil. Along coasts, seaweeds slow erosion by damping wave action. Before the advent of large marine algae, the sea must have been even more potent as a transporting agent than it is today.

Ice, too, wears rock away. Glaciers form where, year after year, winter snowfall exceeds summer melting. At first the snow compacts into beady ice that geologists call névé (skiers call it "corn snow"), then into a fused aggregate of beads, and finally into hard blue ice. When such ice becomes about 30 meters (100 feet) thick (a figure that understandably

Flooding streams and rivers carry many tons of rocks from mountain to valley, often within a few short hours.

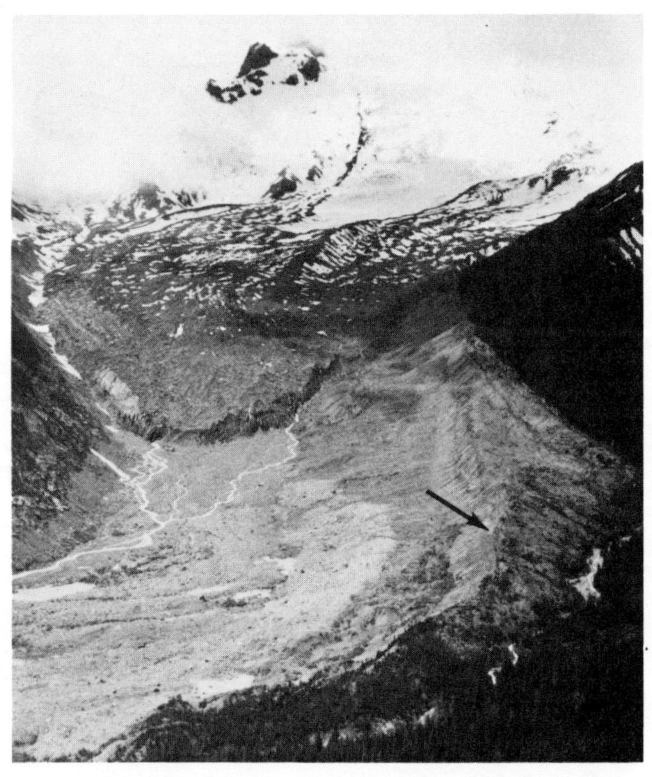

Like other Mount Rainier glaciers, Emmons Glacier (darkened, in this 1980 photo, by Mount St. Helens ash) is now receding. The curve of an old lateral moraine, shown by the arrow, borders a valley partly filled with meltwater deposits known as outwash.

varies with degree of slope), it begins to flow—outward if it is on a flat surface, downhill if it is on a slope.

Ice flows by plastic deformation—constant reorientation of individual molecules, as happens when you knead taffy or Silly Putty. Where flow is rapid, the ice may break and then reassemble, as in the great icefalls of Mount Everest. In mountains, long tongues of ice creep down already existing stream valleys, breaking away rocks loosened by weathering and using them to grind the more solid bedrock beneath. Grinding, scouring, and tearing away, the rivers of ice creep downhill to warmer elevations. Constantly replenished with new snow at the top, constantly melting at their lower ends, glaciers may appear to be standing still; in reality, they are in slow but constant motion. When **alpine glaciers** (also called **valley glaciers**) melt, they leave behind characteristic straightened valleys that are U-shaped in cross section, as well as scoop-shaped **cirques** at valley heads, and large piles of glacier-carried rock, called **moraines,** near their lower ends.

Many theories have been advanced to explain the changes that brought about the great Ice Age glaciers of Pleistocene time. The climate undoubtedly became wetter. Whether or not it initially was much colder is hard to say. Certainly each time that ice sheets covered large parts of America and Europe, air masses blowing across them quickly became quite frigid.

Even without development of glaciers, ice transports rock material. On steep slopes, frost that forms under individual rocks lifts them slightly, nearly always at right angles to the slope. Then, as the frost thaws (at certain times of year a daily event), the rock settles vertically. So in a series of zigzag movements rock and soil material move downhill—a slow process, but like so many geologic processes, long-enduring.

In the related process of soil creep, which may or may not involve frost, downslope transportation is also agonizingly slow. Many years may pass before there is visible change. But change there is, and it is widespread, for soil creep operates on most soil-covered hillsides, characteristically gentling and rounding them.

Though in desert regions it comes into its own, wind is not particularly significant in erosional processes in the region covered in this book. Above

From sharpened peaks to terminal moraines, mountain glaciers leave signs of their passing: U-shaped valleys, cirques, moraines, and outwash gravels.

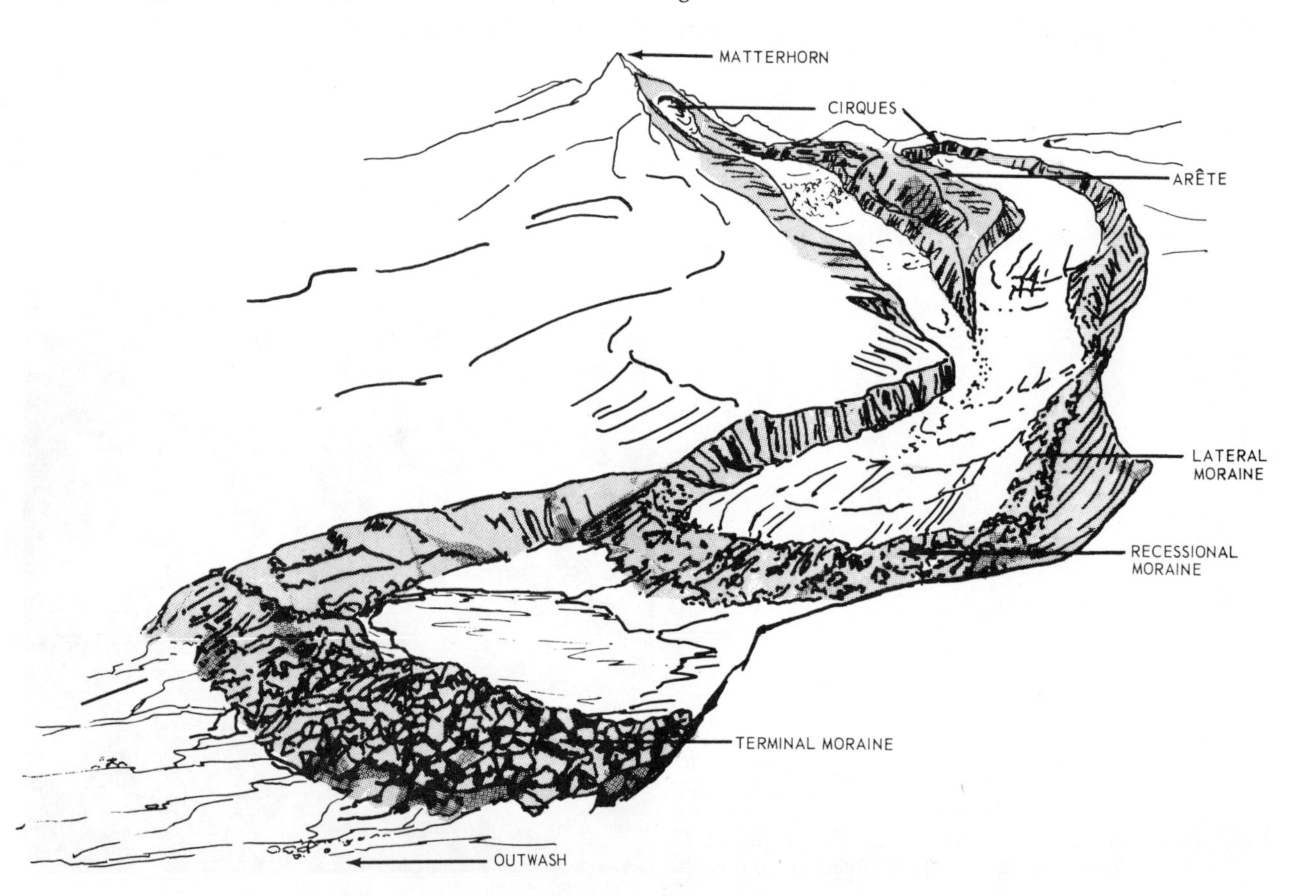

On Mount Rainier, rock fragmented by frost is moved downslope by oft-repeated freezing and thawing.

Trees bent by soil creep can be seen in many western parks.

Along many coasts, waves carve caves and arches. Surf impacting rock loosens individual grains as well as blocks of rock material.

timberline it does help to clean away frost-loosened sand and small rock fragments. And certainly it cleans and rearranges beach sand, swirling loose particles into dunes.

Gravity plays an obvious role in most processes of transportation, controlling the downhill movement of water, ice, and rock. It may act alone, tugging at cliffs until rocks tumble and fall. More frequently, it acts in consort with frost, which loosens the rocks, with snow in avalanches that involve rock material and soil as well as snow, or with water, which lubricates landslides and adds to the weight and mobility of earthflows.

Along coasts, other processes wear the land away. Waves beat with unending fury on rocky headlands, little by little undercutting them, wearing them back. Impacting the headlands with tremendous force, compressing air within every crack and crevice, storm waves literally explode the rocks apart. Charging into hollows, the waves shape clefts, sea caves, and arches which later collapse to isolate cliff-edged islets or **sea stacks.** As they surge back and forth, dragging tools of stone and sand, they carve horizontal benches just at wave level, benches which may extend well out to sea, and which later may be lifted as **marine terraces** well above sea level.

At the same time the waves deposit the eroded debris along bays and other indentations in the coast, as pebbly or sandy beaches. The sea seeks always to straighten the coastline, cutting away here, filling in there. A drive up the Oregon, Washington, or California coast will show to what degree it has succeeded.

V. Understanding Maps and Diagrams

In addition to photographs, this book contains a number of maps, diagrams, and sketches, most of them simplified in order to give a better understanding of the basic geologic patterns in national park areas. Scenically and geologically, the parks and monuments are many-hued. Because they are different from one another—which adds to their attraction, of course—their geology cannot all be illustrated in the same way. This is true of other areas as well, all over the world. Some lend themselves to explanation through **geologic maps,** which show types of rock (often specific formations and their ages) that occur at the surface or just under the loose surface soil and rock debris. Others are better illustrated with **cross sections,** which diagrammatically slice open the crust to give a picture of what geologists have deduced is below the surface. Still others can be explained by combining in **block diagrams** cross sections of the crust with perspective views of surface features, to show what a block of the Earth's crust would look like if it could be lifted out and away from its normal surroundings.

Geologic maps are the prime product of most geologic field research. They represent many hours and days and weeks of slow, patient, often drudging work by geologists or geologic teams who plot outcrop after outcrop, formation after formation, wherever rocks are exposed, to get a coherent picture of the geology of an area (see Olympic National Park in Part 2 for an example). The U.S. Geological Survey has at one time or another geologically mapped most of the United States in considerable detail. Geologic maps of 7½- and 15-minute quadrangles, which cover areas of about 11 × 14 kilometers (7 × 9 miles) and 22 × 28 kilometers (14 × 18 miles), can be obtained from USGS map offices in Reston, Virginia; Denver, Colorado; and Menlo Park, California, as can maps of larger areas—states and, in some cases, individual national parks. Park maps, when they exist, can usually be purchased at visitor centers.

Geologic maps show to the practiced eye not only the rock types but the attitude or position of sedimentary rock layers—whether they are horizontal or tilted, whether they have been folded or faulted, where their contacts lie. Interpretation of geologic maps is a skill well worth learning. The simplified maps in this book are but a beginning; very little detail can be shown at such a small scale. But they will give you at least a rough idea of what to expect beside the road or near the trail as you explore the parks and monuments.

Cross sections and **block diagrams** are usually easier to understand and are good ways to show geologic features at a glance. They are made from maps, or, where rock layers are well exposed, from the rocks themselves. For a sample cross section, see Mount Rainier National Park in Part 2. Often the vertical dimension is exaggerated in cross sections in order to show the succession of rock layers more clearly. Unfortunately, vertical exaggeration also exaggerates folds, steepens faults, and lends unreal—sometimes really startling—ruggedness to the surface profile. Block diagrams, such as those used earlier in this chapter to illustrate folds and faults, are particularly useful for interpreting geologic and topographic changes that come with faulting, folding, uplift, and subsequent erosion. In them, too, the vertical dimension may be exaggerated.

Another type of illustration is the **stratigraphic diagram,** suitable only for stratified (layered) rock. Such diagrams (for an example, see John Day Fossil Beds National Monument) show idealized sequences of layered rocks put together by studying a number of related areas; they depict the strata as they would appear if piled on top of one another once more, in their original horizontal positions. Stratigraphic diagrams may also show how successive rock layers weather—some as slopes, some as cliffs or ledges. Under natural conditions, though, the rock layers may not everywhere show these characteristics; thus one must make allowances in trying to match up scenery and stratigraphic diagrams. The diagrams indicate predominant or *average* thicknesses and *average* weathering characteristics.

In this book the parks and monuments are discussed in alphabetical order. To find your way around in the park areas, refer to the small maps distributed at entry gates by the National Park Service, or purchase more detailed topographic maps at visitor center desks.

VI. Other Reading

In almost every national park and monument visitor center there are books and pamphlets for sale on the geology of that particular area. Some visitor centers have free handout sheets describing local geology. The reverse sides of a few park topographic maps are printed with discussions by U.S. Geological Survey geologists. (A caution: A few of these are geologically out of date.) Nature trail leaflets note geologic features, as do some road guides to park highways.

The books and articles listed below are of a more general nature. Some are college textbooks; others address the geology of whole states or regions, or geologic processes such as volcanism or glaciation. Many of these books are somewhat technical, but others are written for readers with little or no background in geology. All will in some way enrich your understanding of the geology of the parks and monuments.

Alt, D. D., and Hyndman, D. W. 1975. *Roadside Geology of Northern California.* Mountain Press, Missoula, Montana.

Alt, D. D., and Hyndman, D. W. 1978. *Roadside Geology of Oregon.* Mountain Press, Missoula, Montana.

Alt, D. D., and Hyndman, D. W. 1984. *Roadside Geology of Washington.* Mountain Press, Missoula, Montana.

Atkeson, Ray. 1969. *Northwest Heritage: The Cascade Range.* Charles H. Belding, Portland, Oregon.

Axelrod, D. I. 1981. *Role of Volcanism in Climate and Evolution.* Geological Society of America Special Paper 185.

Bailey, E. H. (editor). 1966. *Geology of Northern California.* California Division of Mines and Geology Bulletin 190.

Baldwin, E. M. 1964. *Geology of Oregon.* J. W. Edwards, Ann Arbor, Michigan.

Bullard, F. M. 1962. *Volcanoes: in History, in Theory, in Eruption.* University of Texas Press, Austin, Texas.

Chaney, R. W. 1956. *The Ancient Forests of Oregon.* University of Oregon, Eugene, Oregon.

Chesterman, C. W. 1971. "*Volcanism in California.*" in *California Geology.* California Division of Mines and Geology, vol. 24, no. 8.

Clark. T. H., and Stearn, C. W. 1968. *Geological Evolution of North America.* Ronald Press, New York.

Coffman, J. L., von Hake, C. A., and Stover, C. W. (editors). 1982. *Earthquake History of the United States.* National Oceanic and Atmospheric Administration Publ. 41-1.

Colbert, E. H. 1955 (paperback 1961). *Evolution of the Vertebrates.* Wiley, New York.

Colbert, E. H. (editor). 1976. *Our Continent: a Natural History of North America.* National Geographic Society, Washington, D.C.

Cowen, R. 1975. *History of Life.* McGraw-Hill Book Co., New York.

Crandell, D. R. 1965. "*Glacial History of Western Washington and Oregon.*" in Wright, H. E., and Frey, D. G. *Quaternary of the United States.* Princeton University Press, Princeton, New Jersey.

Decker, Robert and Barbara. 1981. *Volcanoes.* W. H. Freeman and Co., San Francisco.

Dott, R. G., and Batten, R. L. 1981 (3rd edition). *Evolution of the Earth.* McGraw-Hill Book Co., New York.

Easterbrook, Don. J., and Rahm, D.A. 1970. *Landforms of Washington.* Western Washington State College, Bellingham, Washington.

Ekman, Leonard C. 1962. *Scenic Geology of the Pacific Northwest.* Binfords and Mort, Portland, Oregon.

Flint, R. F. 1971. *Glacial and Pleistocene Geology.* Wiley, New York.

Flint, R. F. 1973. *The Earth and its History.* W. W. Norton and Co., New York.

Foxworthy, B. L., and Hill, Mary. 1982. *Volcanic Eruptions of 1980 at Mount St. Helens: The First 100 Days.* U.S. Geological Survey Professional Paper 1249.

Garner, H. F. 1974. *The Origin of Landscapes.* Oxford University Press.

Gilluly, J., Waters, S. C., and Woodford, A. O. 1975. *Principles of Geology.* W. H. Freeman, San Francisco.

Griggs, A. B. 1969. "*Geology of the Cascade Range.*" in *Mineral and Water Resources of Oregon.* Oregon Department of Geology and Mineral Industries Bulletin 64.

Hamblin, W. K. 1975. *The Earth's Dynamic Systems.* Burgess, Minneapolis, Minnesota.

Hamilton, Warren. 1978. *Plate Tectonics and Man.* Reprinted from U.S. Geological Survey Annual Report, Fiscal Year 1976. U.S. Government Printing Office, Washington, D.C.

Harris, Stephen. 1980. *Fire and Ice: the Cascade Volcanoes.* The Mountaineers and Pacific Search Press, Seattle.

Hill, Mary. 1975. *Geology of the Sierra Nevada.* University of California Press, Berkeley, California.

Hinds, N. E. A. 1952. *Evolution of the California Landscape.* California Division of Mines Bulletin 158.

Kay, M., and Colbert, E. H. 1965. *Stratigraphy and Life History.* Wiley, New York.

Kurten, B. 1972. *The Age of Mammals.* Columbia Press, New York.

Macdonald, G.A. 1972. *Volcanoes.* Prentice-Hall, Englewood Cliffs, New Jersey.

Marvin, U. B. 1973. *Continental Drift.* Smithsonian, Washington, D.C.

Match, C. L. 1976. *North America and the Great Ice Age.* McGraw-Hill Book Co., New York.

McKee, Bates. 1972. *Cascadia: the Geologic Evolution of the Pacific Northwest.* McGraw-Hill Book Co., New York.

Oakshott, G. B. 1971. *California's Changing Landscapes: a Guide to Geology of the State.* McGraw-Hill Book Co., New York.

Oakshott, G. B. 1975. *Volcanoes and Earthquakes: Geologic Violence.* McGraw-Hill Book Co., New York.

Post, Austin, and Chapelle, E. R. 1971. *Glacier Ice.* University of Washington Press, Seattle.

Seyfert, C. K., and Sirkin, L. A. 1973. *Earth History and Plate Tectonics, an Introduction to Historical Geology.* Harper and Row, New York.

Sharp, R. P. 1960. *Glaciers.* University of Oregon, Eugene, Oregon.

Stokes, W. L., Judson, S., and Picard, M. D. 1978. *Introduction to Geology: Physical and Historical.* Prentice-Hall, Englewood Cliffs, New Jersey.

Sullivan, W. 1974. *Continents in Motion.* McGraw-Hill Book Co., New York.

Vine, F. J. 1970. *Sea-Floor Spreading and Continental Drift.* Journal of Geological Education, vol. 18, no. 2.

Whitney, Stephen. 1979. *The Sierra Nevada.* Sierra Club Books, San Francisco.

Whitney, Stephen. 1983. *A Field Guide to the Cascades and Olympics.* The Mountaineers, Seattle.

Williams, Howel. 1952. *Ancient Volcanoes of Oregon.* Condon Lectures, University of Oregon Press, Eugene, Oregon.

Wilson, J. T., and others. 1972. *"Continents Adrift."* Readings from *Scientific American.* W. H. Freeman and Company, San Francisco.

Wyllie, P. J. 1971. *The Dynamic Earth.* Wiley, New York.

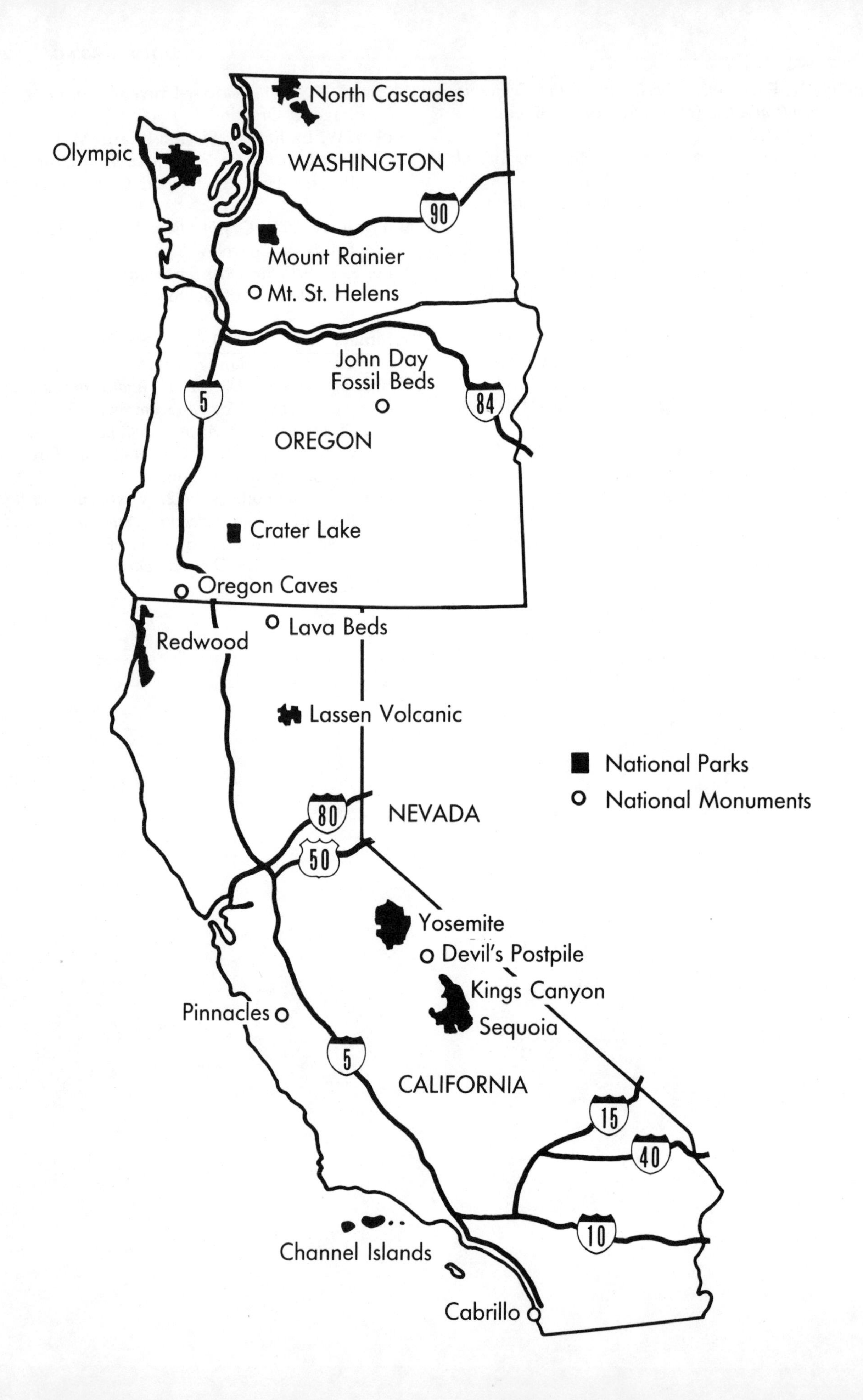
North Cascades
Olympic
WASHINGTON
90
Mount Rainier
Mt. St. Helens
John Day
Fossil Beds
5
84
OREGON
Crater Lake
Oregon Caves
Lava Beds
Redwood
Lassen Volcanic
National Parks
National Monuments
80
NEVADA
50
Yosemite
Devil's Postpile
Kings Canyon
Sequoia
Pinnacles
5
CALIFORNIA
15
40
10
Channel Islands
Cabrillo

PART 2.

THE NATIONAL PARKS AND MONUMENTS

Cabrillo National Monument

Established: 1913
Size: 0.6 square kilometers (0.22 square miles)
Elevation: 116 meters (380 feet) at visitor center, 130 meters (428 feet) at old lighthouse
Address: P.O. Box 6670, San Diego, California 92106

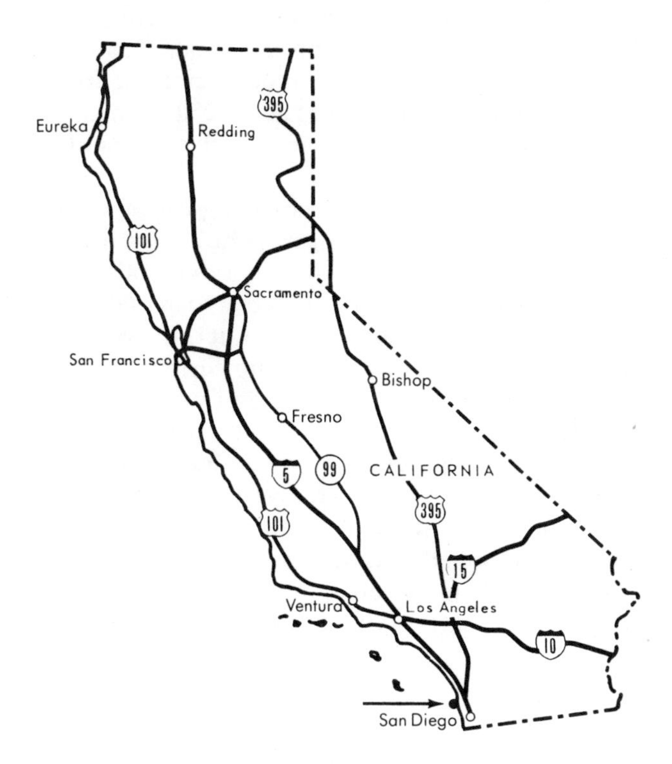

STAR FEATURES

• A cliff-edged promontory that played a significant part in development of San Diego Bay.

• Cretaceous sandstone, shale, and conglomerate, marine deposits formed as mainland ranges rose.

• Examples of coastal processes such as sea-cliff erosion and development of sandspits, submarine canyons, and marine terraces.

• An overview of San Diego Bay, the sandspit that protects it, and mainland geologic features.

• Visitor center, naturalist talks, a nature trail, and a road down to the wave-cut bench with its many interesting tidepools.

SETTING THE STAGE

At the very tip of Point Loma, this national monument commemorates explorer Juan Rodriguez Cabrillo, who sailed into San Diego Bay in 1542. In addition to its historical attraction, the area offers a glimpse of prehistory in the geologic features of Point Loma, as well as in views of San Diego and San Diego Bay.

Point Loma itself, a slender fault block lifted high above neighboring fault blocks, is made up of marine sandstone, shale, and conglomerate. These sedimentary rocks are found elsewhere as scattered remnants along the southern California coast. They were deposited in Cretaceous time in shallow sea waters just west of the westward-drifting continent. Some of them, recognized by grading of individual beds from coarse-grained to fine-grained, are turbidity deposits, sand and mud released rapidly by submarine landslides near the edge of the continental shelf.

The high, irregular ridge of Point Loma, with its east-dipping strata, is the east limb of an anticline that also dates back to Cretaceous time. The western side of the anticline is thought to lie 6 kilometers (4 miles) out to sea. The Point Loma ridge shelters the northern end of San Diego Bay from prevailing west and northwest wind and waves, and establishes the counterclockwise circulation of nearshore waters that is responsible for the protected bay, the largest natural harbor south of San Francisco. The circulating water picks up sand at the mouth of the Tijuana River, just north of the Mexican border, and carries it northward, depositing it as a long sandspit—the Silver Strand—that connects the town of Coronado and North Island to the mainland and defines San Diego Bay.

Like other streams flowing to the sea in this region, the Tijuana River has now been dammed.

Much of the sand that would normally reach the sea remains trapped in reservoirs behind the dams. As a result, sand lost during winter storms is not fully replaced by summer currents, and beaches and sandspits along the coast are becoming narrower. The dams also prevent enlargement of the harbor's protective spit, which eventually, if left in its natural state, might connect with Point Loma, sealing off the harbor. Problems of dredging the bay and maintaining the channel should in theory be helped by the damming of the river.

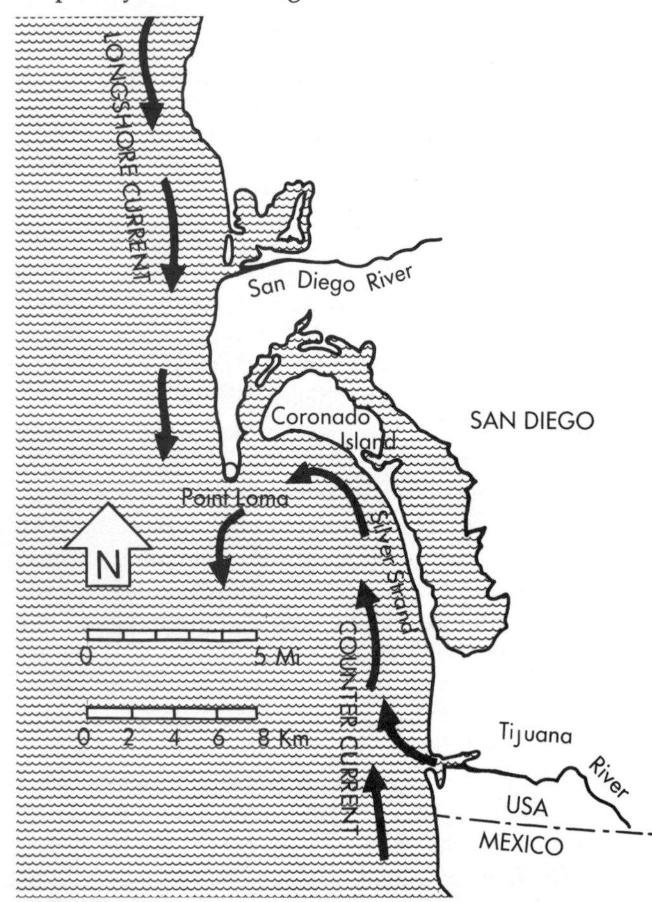

Jutting westward from the California coast, Point Loma sets up a countercurrent that carries sand north from the Tijuana River to build the long Silver Strand bar, sheltering San Diego's harbor.

The city of San Diego and most of its suburbs lie on a series of old wave-cut marine terraces that slope gently toward the Pacific. Three terrace levels are recognized in this area: the La Jolla Terrace roughly 20 meters (60 feet) above sea level, well developed near Mission Bay north of Point Loma; the Linda Vista Terrace between 100 and 150 meters (300–500 feet) above sea level, on which most of metropolitan San Diego is located; and remnants of the Poway Terrace some 250 to 350 meters (800 to 1200 feet) in elevation.

Marine terraces are sea-cut platforms later raised above sea level. Initially they formed at wave base, a few feet below the actual surface of the sea, where wave erosion is most effective. Terraces here and elsewhere along the California coast are now above sea level because of uplift along faults. Faults also offset different parts of each terrace. Most of the faults parallel the coastline. Easily visible from Point Loma, a recently active fault runs along the base of the bluffs behind the northern part of the city, connecting the center of the city through Rose Canyon (route of Interstate 5) to the coast north of La Jolla.

Much of the fault movement in this area, as well as in the rest of coastal California, is horizontal rather than vertical. The horizontal movement parallels the coast, as seaward segments move north relative to landward segments. This is the same type of horizontal offset that occurs along the famous San Andreas Fault, far inland east of San Diego. Onshore faults have been recognized for some time; offshore faults, only recently discovered, create an unusual continental border of submarine ridges, shallow submarine platforms, and long, deep basins paralleling the coast. This rugged, mountainous border, hidden of course from the eyes of man, extends out about 250 kilometers (150 miles) from shore (see map with Channel Islands section).

Also hidden beneath the sparkling Pacific is the La Jolla submarine canyon, at right angles to the coast. It cuts through the offshore ridges and platforms west of La Jolla. One of several such undersea canyons along the California coast, it is the work of a sand- and mud-laden longshore current coming from the northwest, turned aside by the La Jolla headland. Dense and heavy with its sediment load, the current flows like a river down the continental slope, scouring an irregular, ever-deepening channel. It ultimately drops its sand and silt on a submarine fan at the canyon's mouth, about 40 kilometers (25 miles) offshore, where the water is around 1000 meters (3300 feet) deep.

GEOLOGIC HISTORY

The story of the San Diego area begins in Jurassic time with intrusion of the granite that makes up the core of the Peninsular Ranges east of San Diego. The granite is of the same age and origin as some parts of the Sierra Nevada's Batholith (see Part I and Yosemite National Park). In Cretaceous time, sandstone, shale, and conglomerate—the rocks now exposed on Point Loma—were deposited in shallow seas that bordered the continent. Along with these marine sediments are some volcanic rocks, relics of long-gone volcanoes erupting along the subduction zone where the Pacific Plate dives beneath the edge of the continent. The southwest-

National Park Service photo

Sea cliffs expose tilted rocks of turbidity deposits, beveled and overlain by thick sandstone of the Cabrillo Formation. Parts of a wave-cut terrace lie at the unconformity (arrow) between the two formations. Note the many rills marking the soft sandstone—products of rain and sea spray.

National Park Service photo

Monument visitors examine the layered turbidite *of the Point Loma Formation. In each unit, a few inches thick, sand at the bottom grades upward into silt and clay, in a pattern characteristic of sediments formed as grains of different size, suspended in muddy (turbid) currents, settle to the sea floor.*

ward movement of this part of the continent in late Cretaceous and early Tertiary time caused a certain amount of warping and bending, and is responsible for the broad anticline of which Point Loma is the eastern part.

In Tertiary time the continent—the North American Plate—continued its southwestward drift and its overriding of the East Pacific Plate, and more sedimentary and volcanic rocks were deposited along the coast. Uplift and westward tilting elevated the Peninsular Ranges, and vertical and horizontal movement along many faults gradually brought the coastal area close to its present configuration.

Periodic earthquakes in this area show that movement along the faults goes on today. Historical records indicate that there has not been a really severe quake—of magnitude greater than 4.0 on the Richter scale—in historic time. In 1964, though, three sharp tremors of magnitude 3.7, 3.6, and 3.5 resulted from sudden movement along the part of the Rose Canyon Fault that extends south beneath San Diego Bay.

The sea is responsible, as we have seen, for the final shaping of the coast, as rivers and streams are for details of mainland geography. With periodic uplift and, in Pleistocene time, periodic drops in sea level as great ice sheets farther north tied up sizeable volumes of water, the pounding waves carved a series of nearly level terraces, each of them partly cut away as the next one formed. The La Jolla and Point Loma regions, however, remained as seaward-jutting headlands. South of Point Loma, a counterclockwise current became established, gradually to build up the sandspit that now shelters San Diego Bay.

Along the seaward side of the Point Loma headland, another marine platform is now being shaped by the sea. Able to penetrate the wave-damping effects of thick offshore kelp "forests," Pacific storm waves still strike with force, hollowing out channels, caves, tunnels, and tidepools in Cretaceous rocks of the headland. As the steep bluffs are undermined, they collapse. Piles of landslide rubble are eventually swept away by the sea. Broken into sand grains and rounded pebbles, they will be carried along the shore and deposited as beaches or sandspits. Or they may become part of the turbidity currents that flow down submarine canyons, in which case they will come to rest in submarine fans far from shore.

OTHER READING

Abbott, P. L., et al. 1979. *Geological Excursions in Southern California Area,* San Diego State University.

Pryde, P. R. 1976. *San Diego: An Introduction to the Region*. Kendall/Hunt Publishing Company.

Channel Islands National Park

Established: 1938 as a national monument, 1980 as a national park
Size: 1009 square kilometers (390 square miles)
Elevation: 0 to 747 meters (0 to 2450 feet)
Address: 1699 Anchors Way Drive, Ventura, California 93003

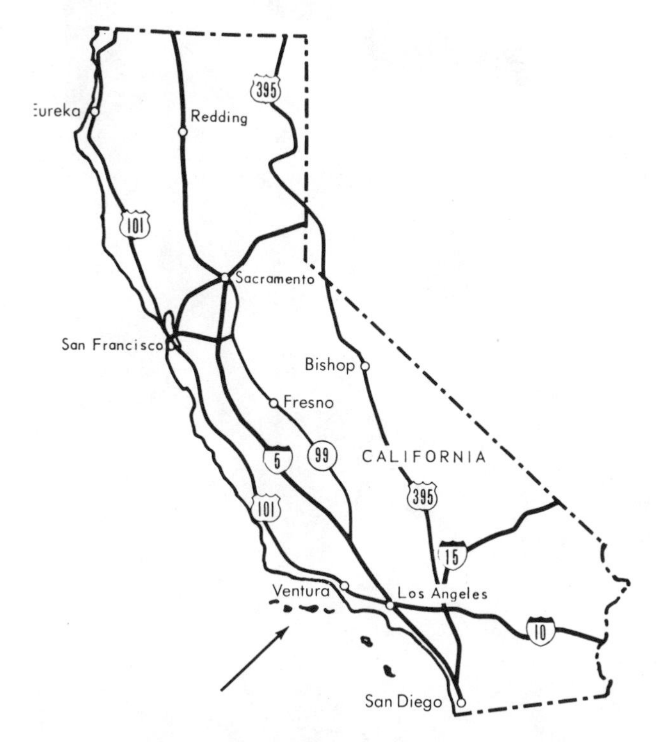

STAR FEATURES

• Mountains foundered in the sea, with only their highest peaks now showing above the waves.
• Jurassic intrusive rocks related to those in the Sierra Nevada, as well as Cretaceous and Tertiary sedimentary rocks and Tertiary volcanic rocks.
• Evidence of northwestward movement along faults that parallel the San Andreas Fault of the mainland.
• Prime examples of shore processes, with beaches, cliffs, sea caves, sea stacks, and gradual wearing away of the islands by pounding surf and surging currents.
• Visitor centers (mainland and Anacapa) and guided walks (Anacapa).

SETTING THE STAGE

The Channel Islands, offshore extensions of the Santa Monica Mountains, are geologically akin to the mainland, to the clustered east-west ridges known as the Transverse Ranges. Their intrusive rocks seem to be related to the massive intrusions of the great Sierra Nevada Batholith, as well as to smaller intrusions farther south. Their sedimentary and volcanic rocks, also kin to those of the mainland, document the evolution of the California coast during Cretaceous and Tertiary time.

With the rest of coastal southern California, the Channel Islands are drifting northwest a few inches

Sea caves, constantly reamed by surging waves, developed along faults, joints, and other weakened zones in the rock.

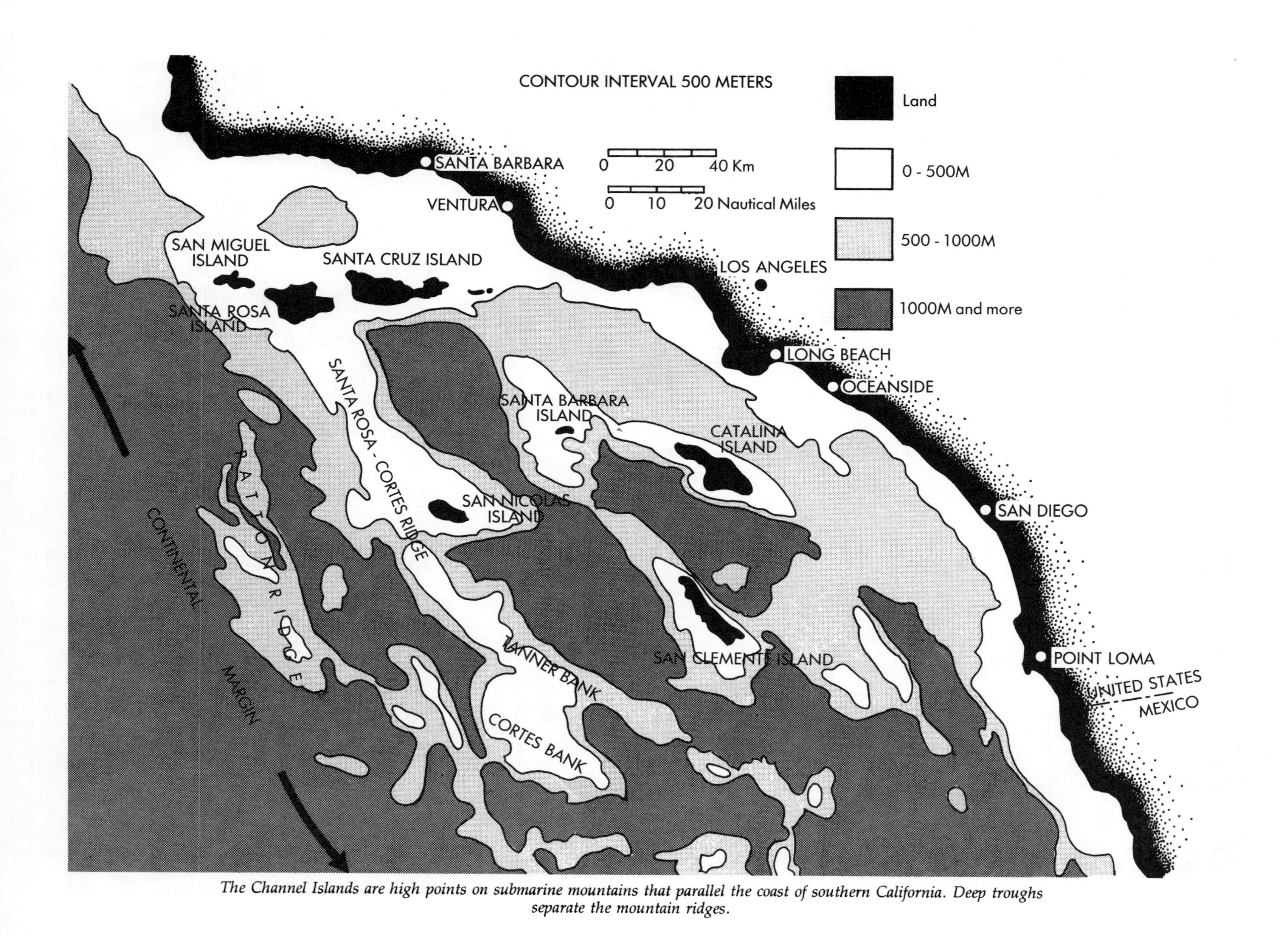

The Channel Islands are high points on submarine mountains that parallel the coast of southern California. Deep troughs separate the mountain ridges.

a year—a jerky, spasmodic drift that manifests itself in periodic earthquakes. Though the San Andreas Fault lies 50 to 65 kilometers (30 to 40 miles) inland, there are literally hundreds of other faults between it and the coast, and more hundreds on the sea floor between the mainland and the Channel Islands. A number of the faults cut through the islands themselves.

The national park includes the four northernmost islands—San Miguel, Santa Rosa, Santa Cruz, and little Anacapa—as well as Santa Barbara Island farther southeast. Anacapa is the most frequently visited, with daily tour boats in summer.

San Miguel and Santa Rosa Islands, the westernmost of the group, connect with the mainland by a submarine ridge that extends southward from Point Conception on the California coast. This undersea ridge extends on to San Nicholas Island some 95 kilometers (60 miles) farther southeast. It intensifies upwelling of deep, cold, nutrient-rich waters all along this part of the coast, bringing about the abundance of sea life for which the islands are famed.

San Miguel, with an area of about 40 square kilometers (16 square miles) is a rugged, double-summited island with hilly uplands cut by deep ravines. Exposed to the full force of the sea and to waves driving in before the prevailing northwest winds, its coast is sharp and rugged, with many offshore islets. Wind erosion, in part the result of overgrazing, has formed cup-shaped blowouts in various parts of the island. With a plentiful supply of sand, dunes have developed as well. An interesting feature of the island is the "caliche forest" of lime-saturated tree trunks—a fossil conifer forest.

Santa Rosa is larger than San Miguel—about 223 square kilometers (86 square miles). Its long beaches and rocky bluffs are backed by gentle grass-covered hills. Like San Miguel, it contains a caliche forest; fossilized bones of a small mammoth

Broad bands of kelp shelter island cliffs from the full force of Pacific waves. They serve as homes for the myriad marine animals that make the islands attractive to scuba divers and snorkelers.

have also been found on the island, though how the beast got there nobody knows.

Santa Cruz is the largest of the national park islands, 250 square kilometers (97 square miles). The deep wooded canyons and steep cliffs of its north shore and the cliffs, beaches, and coves of its south shore border a hilly interior crowned with summits ranging to 732 meters (2400 feet). Long east-west summit ridges are separated by a slender valley, all products of movement along the Santa Cruz Island Fault, which runs the length of the island. The shoreline is wild and delightfully unaltered by man, with deep canyons, towering cliffs, and sea caves interspersed with beaches and coves.

Anacapa is tiny—three small islands, the sum of whose land is less than 3 square kilometers (1 square mile). It is surrounded by formidable cliffs eroded by the sea, with caves and steep-walled coves. A tilting plateau tops the eastern two islands; on the westernmost, cliff meets cliff in a knifelike ridge.

Santa Barbara, also quite small, is similarly composed of volcanic rock, and has a steep, rugged

Six wave-cut terraces mark the west end of Anacapa Island. Each represents a still-stand in uplift of the island.

Once attached to Anacapa Island, this sea stack still bears a flat upper surface, formerly part of a marine terrace. The arch may well have started out as a sea cave.

shoreline punctuated by a few beaches.

Both San Miguel and Santa Rosa show geologic kinship with the Santa Ynez Mountains on the mainland near Point Conception. Like the Santa Ynez Mountains they are composed of sedimentary rocks—conglomerate, sandstone, and shale. The shales in particular have interested petroleum geologists: They are rich in animal and plant material and are probably the source-rocks for some of the petroleum wealth of this area.

Santa Cruz, Anacapa, and Santa Barbara are made up of volcanic rocks, with a much smaller representation of sedimentary rocks. Some 2500 meters (8000 feet) of lava flows, measured perpendicularly to their original horizontal position, appear on Santa Cruz. Much of the lava there, as well as that on Anacapa and Santa Barbara, is pillow lava, known to have erupted beneath the sea. In its varied geologic makeup Santa Cruz Island resembles the Santa Monica Mountains on the mainland. It does have some rock types unique to the islands: the Santa Cruz Island Schist, which forms a long east-west ridge almost the length of the island, and a parallel sliver of granite. These rocks lie just south of the Santa Cruz Island Fault, which also runs the length of the island. The granite is similar in composition and appearance to that of the Sierra Nevada and the Peninsular ranges farther south.

Once the tops of mainland mountains, the islands were separated from the mainland by movement along faults, many of them well known because of the oil-rich nature of this region. Here, fine-scale geophysical studies have been made of the sea floor itself: precise echo sounding, seismic studies that bounce shock waves off the bottom, and measurements of changes in gravity across the area. Because of its oil wealth, the region between the islands and the mainland is one of the most thoroughly studied sea-floor areas in the world. It is not at all a typical continental shelf, but rather an irregular surface of elevated blocks and ridges separated by deep basins. Total relief between the ridges and the rock floors of the basins reaches about 2600 meters (8500 feet). Many of the basins are partly filled now with soft, unconsolidated marine sediments.

Most of the faults in this region, both on the mainland and on the floor of the Santa Barbara Channel, follow the nearly east-west alignment of the coast. Though many faults account for the vertical dropping of the sea floor itself, and for the differences in elevation of the sea-floor ridges and basins, movement along others is nearly horizontal, with the seaward side moving west or northwest relative to the landward side. Santa Cruz Island provides a good example of this horizontal movement: The portion of the island that lies south of its prominent lengthwise fault has moved many miles westward relative to the northern part of the island, and shows rocks that are similar to those in the mountains near Los Angeles.

Marine erosion has created many interesting features on these islands, from their sea cliffs and terraced slopes to sea caves, arches, and offshore stacks. The large kelp "forests" that now surround these islands reduce erosion by damping wave action, but still the sea crashes against the cliffs, compressing air into crevices large and small, blasting away bits and pieces of rock and gradually reshaping the islands, destroying them little by little and piece by piece.

GEOLOGIC HISTORY

Mesozoic Era. The oldest rocks exposed in Channel Islands National Park—the granite and schist along the crest of Santa Cruz Island—are thought to be of Jurassic age. Little is known about the granite except that it is similar to the immense intrusions of the Sierra Nevada, as well as to other intrusions farther south in the Peninsular Ranges near San Diego.

Cretaceous rocks, too, are few and far between—thick layers of marine sandstone and shale on San Miguel Island, and similar rocks known from cores removed from wells on Santa Cruz Island. Their counterparts on the mainland are coarser sandstone and conglomerate, deposited on land rather than in the sea. They tell us that the shelving Cretaceous shoreline—the west edge of the new North American continent—was not far from the present California coast.

Cenozoic Era. Marine conditions continued well into Tertiary time, with thick deposits of Paleocene, Eocene, and Oligocene sedimentary rocks known today from exposures on San Miguel, Santa Rosa, and Santa Cruz Islands. Mainland counterparts of these rocks include the Sespe Formation, a series of easily recognized, coarse, red-brown floodplain deposits that show us once more that the shoreline was at that time quite nearby. A few outcrops of the Sespe Formation also occur on Santa Rosa Island.

By Miocene time the picture had changed. Several long, narrow marine basins formed, troughlike depressions that deepened with time. In these troughs, which extended onto the present mainland near Santa Monica, thousands of meters of sedimentary rocks were deposited, ranging from the coarse conglomerate of undersea landslides to fine mudstone and limestone and cherty, porcelain-like shale. Fine volcanic ash settled to the sea floor from volcanic eruptions on the mainland, contributing the silica that hardens the porcelain shale. More silica was contributed by tiny silica-shelled plants called diatoms, which lived in well lit surface waters but settled to the sea bottom as they died. With time, the rich supply of organic material from the diatoms and from other tiny marine organisms became oil—petroleum—that migrated into surrounding coarser sediments, now the oil and natural gas reservoirs tapped by onshore and offshore wells.

Also in Miocene time, sea-floor eruptions created thick masses of pillow lava, some of which appear on the islands today. Lava erupting under water forms bulging lobes and cools rapidly, crusting and solidifying into pillow-shaped blobs a foot or so across. Successive pillows form as the lava continues to erupt, piling up on the sea floor. Pillow basalt is visible today on several of the Channel Islands, Anacapa among them.

Many changes came to this region in Pliocene and early Pleistocene time, changes related to faulting and folding as the continent continued to drift west, and to fluctuation of sea level during the Ice Ages. The mountains of the Transverse Ranges came into being as the land bent and broke in long, fault-edged anticlines. Offshore faulting thrust the Channel Islands above the sea, probably as a much more continuous ridge than exists today, and created the deep Santa Barbara Channel between the islands and the mainland. At the same time, a shearing movement began on many faults as the coastal islands were dragged westward relative to the shore. Some of the sedimentary rocks now exposed on Santa Rosa, Santa Cruz, and Anacapa Islands were thus carried west from the Los Angeles Basin. Santa Cruz Island itself consists of two narrow fault slivers, the southern one moving west relative to the northern one. Repeated movement along many of the faults continues to this day.

Pleistocene sea level changes were brought about as vast continental ice sheets accumulated, tying up water that normally would return to the sea. The shifting sea pounded away at new borderlands, reshaping both the Channel Islands and the mainland. Since erosion proceeds most rapidly within the surf zone, new marine terraces were planed off at various levels. Later uplift tilted these terraces and raised them well above sea level. They are evident now at the tops of sea cliffs on the islands and along the mainland coast.

OTHER READING

Howorth, P. C. 1982. *Channel Islands: the Story behind the Scenery*. KC Publications, Las Vegas, Nevada.

Crater Lake National Park

Established: 1902
Size: 744 square kilometers (286 square miles)
Elevation: 1,438 to 2,721 meters (4,745 to 8,926 feet)
Lake surface elevation: 1,882 meters (6,176 feet)
Address: Crater Lake, Oregon 97604

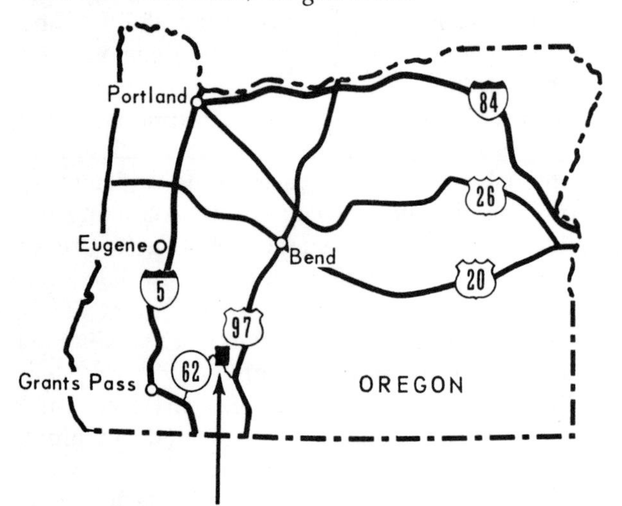

STAR FEATURES

• Lovely Crater Lake, reflecting in its peacock waters the bare and rocky walls that encircle it—one of the world's youngest, least eroded, and most easily studied volcanic calderas.

• Other volcanic features, including unusual pinnacles of pumice thought to have hardened around clustered steam vents.

• Glacial features of a once majestic mountain: glacial striae and U-shaped valleys beheaded by the volcano's climax convulsion.

• A leisurely and informative boat tour around the inside of a former volcano.

• Visitor center exhibits, conducted hikes, daytime and evening talks, numerous short trails (some with nature leaflets), and a 53-kilometer (33-mile) Rim Drive, with numerous turnouts to view the lake and geologic features of the park.

See color section for additional photographs.

SETTING THE STAGE

Superimposed on older volcanic rocks, the long line of the Cascade volcanoes stretches 950 kilometers (600 miles) from Mount Garibaldi in the Coast Mountains of southern British Columbia to Lassen Peak in northern California. The volcanic peaks, not all of them identical in age, include Mount Rainier, Mount Hood, Mount Shasta, and Mount St. Helens, whose destructive 1980 eruptions drew the attention of geologists and the world at large. They also once included a volcano we now call Mount Mazama, probably for a time one of the highest peaks in Oregon.

Mount Mazama is here no longer. In its place, low on the skyline, is the rim of the beautiful lake that some have called the "Gem of the Cascades."

Early studies of Crater Lake, about a century ago, convinced geologists that Mount Mazama had once existed and that only its base and lower slopes now remain. On the ring of cliffs surrounding Crater Lake can be read the story of the ancient mountain. Irregular, alternating layers of lava and volcanic ash tell us the nature of the volcanic outpourings that shaped the mountain and show us that, like some other Cascade giants, it was a cluster of overlapping stratovolcanoes, perhaps with one cratered summit high above the lesser summits of several satellite companions. Resistant vertical dikes that once tapped the subterranean magma supply were feeder channels for some of the smaller satellite cones. Striated rocks on the present rim, and U-shaped valleys shown in cross section along the southeastern rim, tell us of several glacial episodes, adding evidence that the upper slopes of the great volcano were once high enough to be heavily sheathed in snow and ice.

What, then, happened to this volcano? Early guesses were that an explosion of cataclysmic proportion shattered and blasted away the upper two-thirds of the mountain. Unfortunately, a search of the surrounding area turned up few rock fragments that might have been parts of the former mountain summit, certainly not enough of them to account for the whole of the lost mountain mass, which was calculated to exceed 50 cubic kilometers (13 cubic miles). Instead, the search revealed more and more fine yellow pumice and ash scattered far and wide to the northeast—to Yellowstone National Park in Wyoming and in lesser quantity to Alberta, Saskatchewan, and the Dakotas, an area of about 900,000 square kilometers, or 350,000 square miles.

Over the years, continuing research introduced a new interpretation: In its final, explosive eruption, Mount Mazama emitted such vast quantities of frothy pumice and lava that it brought about its own collapse.

The crystal waters of Crater Lake are derived solely from rainfall and melting snow. Their deep color can be attributed to the clarity of the lake water and to its depth—nearly 600 meters (2,000 feet). The amount of water added each year seems to be in balance with that lost by evaporation and by slight leaking through the walls. The level varies only a little with the seasons. Except for the leaking, there is no natural flushing mechanism here as there is in lakes fed and drained by streams and rivers. To preserve the pristine clarity of this lake, the Park Service asks visitors to refrain from leaving wastes or debris of any kind in or near its waters. The deep water rarely freezes over in winter—only once in historic time. Is it warmed from below, from the embers of Mount Mazama's fires? Recent measurements have turned up tiny temperature discrepancies at the bottom, where in some localized areas there seem to be underwater hot springs on the floor of the lake.

Volcanic rocks around Crater Lake are mostly andesite and dacite, two types of gray volcanic rock composed largely of plagioclase (a feldspar mineral) with minor additions of quartz and dark minerals. These rocks can and do occur as lava, bubble-filled scoria, broken and recemented breccia, pebblelike cinder, and fine volcanic ash. Basalt, darker and with more iron minerals, occurs at Red Cone (where the iron is oxidized to a brick red color), Desert Cone, Timber Crater, Bald Peak, and Union Peak.

In outlying parts of the park are a number of other volcanic features that survived Mount Mazama's destruction or developed after it. They are discussed individually later.

GEOLOGIC HISTORY

Cenozoic Era. Crater Lake's story began about 65 million years ago, as the North American Plate broke free and drifted westward, riding up over the oceanic East Pacific Plate. Under the long-continued strain of the collision between plates, the crust broke and buckled, and eventually—in Miocene or Pliocene time—the Cascade Range was formed. Close to the crest of these new mountains, basalt eruptions built low-profile volcanoes that coalesced into a broad volcanic plateau.

Within the last 3 million years, eruptions along the crest of the Cascades gradually changed their pattern. The lavas themselves became lighter in color, less fluid, and more explosive. They tended to pile up in alternating layers of lava and ash, as individual steep-sided cones that towered over and buried the earlier lava plateau. But even as the new volcanoes grew in height and mass, erosion struck—the product of glacial ice and torrential mountain rains and snows.

The Cascade volcanoes did not all develop at once. Some were old and deeply eroded by the time that Mount Mazama, a relative youngster, was born in southern Oregon only 200,000 or 300,000 years ago. This was a mountain not of a single volcanic cone but of a cluster of cones that drew their sustenance from magma chambers far below.

As the clustered volcanoes increased in size, lava flow by lava flow and ash fall by ash fall, the central summit became high enough that glaciers again and again sheathed its summit only to be destroyed by new eruptions. By 7,000 years ago (a date derived from measurement of radioactive carbon in charred wood embedded in ash flows) Mount Mazama, once 3,600 meters (12,000 feet) high, must have been deeply eroded. By 7,000 years ago, glaciers had retreated from the valleys, and a necklace of satellite cones had developed along circular fissures that partly ringed the central peak.

Seen close up, the layered caldera walls contain thick masses of breccia like that at the right. Most of the breccia formed atop moving lava flows as cooling lava cracked and broke. Phantom Ship is in the distance.

Then, deep underground in the great feeding chamber, magma pushed upward. This, too, had happened in the past. But for some reason there was a difference now, and when some of the magma broke through to the surface, Mount Mazama's doom was sealed.

1. In Mount Mazama's climax eruption, avalanches of incandescent ash sped down the mountainside while darkening clouds rained pumice on the slopes below.

2. During its climax eruption Mount Mazama collapsed into its partly emptied magma chamber, leaving remnants of climactic ash on the rim of a giant caldera.

3. The caldera slowly filled with rain and melted snow, and Wizard Island, a cinder cone with lava flows at its base appeared near the western wall. Another small volcano is hidden by the lake.

Drawings adapted from paintings by Paul Rockwood

The tawny summit of Redcloud Cliff displays remnants of Mount Mazama's last eruption: light-colored pumice from the beginning of the climax eruption. This view shows the long rockslides that plunge toward and into the lake.

In its underground chamber, magma may separate into heavy, iron-rich, partly crystallized fluid that tends to sink, and far lighter, gas-rich fluid that tends to rise—the same type of separation that occurs when a pot boils over. Then, anything that initiates an eruption—an earthquake, movement of the magma, upward melting of crustal material, landslides occasioned as mountain slopes swell above the magma's thrust—will lead to expulsion of some of the gas-rich upper part of the fluid, which explodes in gargantuan incandescent bursts of froth. The result of such effervescence is pumice, a rock that looks almost like Styrofoam and that is so air-filled that it floats on water. A decrease in subterranean pressure as a result of the first few bursts simply amplifies the eruption. Gas in the remaining froth effervesces like uncorked champagne, and it too spurts through the volcanic vent, allowing still more gas to expand.

As eruptions at Mount Mazama became ever fiercer, earthshaking explosions rocked the mountain. (The date was about 4860 B.C.; man roamed Africa, Eurasia, and the Americas but had built no cities, written no words.) Tremendous billowing ash clouds blackened the sky, reached into the stratosphere and drifted northeastward on the wind. Giant boulders were tossed like pebbles from the volcano's maw. Red-hot ash avalanched at hurricane speeds down its slopes, carrying with it volcanic bombs and blocks of pumice and rock derived from deep in the Earth's crust. Pumice rained onto the mountain slopes. Snow melted and vaporized. Forests—those left standing—burst into flames. (Sound familiar? We've had but a tiny taste in Mount St. Helens' 1980 eruptions, minute by comparison with Mount Mazama's grand finale.)

Perhaps after many successive explosions, or perhaps after just one, the volcanic thunder died away. The ash clouds dissipated, revealing a scene of utter desolation. Nothing moved save plumes of acid steam rising from the ash. Nothing lived. From southwestern Oregon to Saskatchewan, suffocating gray-white ash blanketed forest and prairie. Close to the ravaged mountain, a moonscape in gray and white.

But underneath the mountain, in the magma chamber, the level of molten rock had been lowered. Cracked and broken by its ordeal, undermined by the removal of supporting magma, Mount Mazama itself sank to fill the void. Little of the older mountain had been destroyed in the great eruption; it all collapsed into the depleted magma chamber. Geologists believe that the end came rapidly, perhaps in a few days' time. Circular fissures simply widened and lengthened, and the crippled central cone slid downward, piecemeal or all at once, leaving where it had stood a steep-walled, circular caldera 1,200 meters (4,000 feet) deep. When first formed, the caldera was an empty,

Although they emerge at almost the same elevation as the lake surface, Annie Spring's clear waters come from rain and snowmelt, not from leakage from Crater Lake.

gaping chasm devoid of life, steaming with volcanic gases, possibly pooled with lava.

Since then, the volcano has rested—but not died. Massive new flows smoothed the rough and broken floor, a volcanic dome may have formed, and a small cinder cone built up in the northeastern part of the caldera. And then another, in the western part—the cone that is now Wizard Island. A small dome later pushed through its east flank, and more lava flows spread from its base. But there were no more catastrophic eruptions. Rain and snow falling in the caldera accumulated there, and a shallow lake was born, a lake that gradually deepened to nearly 600 meters (2,000 feet)—its present depth.

Will Mount Mazama erupt again? We do not know. Is this sparkling lake to be destroyed, a transient moment of elegance in a lifetime of fury? Eventually yes, by erosion if not by further volcanism.

BEHIND THE SCENES

Crater Lake boat trip. Only by taking the boat trip can you sense the true size of Crater Lake and the height of its sheer walls. The tiny boats seen from the rim are sizeable open launches that carry 60 passengers. They dock at Cleetwood Cove in the northeastern part of the Crater Lake caldera, at the base of a trail leading down from the rim. (The trail is steep, but the boat trip is well worth the effort.)

The boats circle the lake, introducing their passengers to the inner workings of Mount Mazama. A Park Service interpreter explains the geologic features, which include alternating layers of lava flows and volcanic ash emitted during Mount Mazama's early growth, dikes that fed satellite volcanoes, and several of the satellite volcanoes themselves, sliced in half by their parent's collapse. The great rockslides below the Watchman, Dutton Cliffs, and Red Cloud Cliffs are seen also, as are Castle Rock, made of pumice welded together by its own heat, and Wizard Island, a volcano within a volcano.

Devils Backbone. The most prominent of about 20 dikes easily visible on Crater Lake's cliffs, Devils Backbone can be seen from the rim or the boat tour. It formed when molten rock slowly cooled in one of many fissures radiating from Mount Mazama's central conduit. The rock is of the same composition as most of Mazama's lava flows. Its relative hard-

Devils Backbone stands out from the caldera wall because its rock is more resistant than adjacent ash and lava flows.

ness has led to its present prominence; softer ash layers and jointed lava flows have eroded away from around it.

Garfield Peak (elevation 2,457 meters, or 8,060 feet). The summit of this peak is tipped with light-colored pumice from Mount Mazama's final blast. The trail to its summit will give you excellent vantage points for viewing Crater Lake and the surrounding country. To the north are Mount Thielson and Mount Bailey, deeply eroded Cascade volcanoes older than Mount Mazama. To the south are Mount McLoughlin, and Mount Shasta with its young companion Shastina. Along the trail are good exposures of volcanic breccia, full of irregular lumps of volcanic rock. On the cliff side, the breccia is interlayered with lava flows, a feature that shows up better from other rim viewpoints or from the lake surface. The breccia itself may have formed atop moving, cooling lava flows.

Hillman Peak (elevation 2,486 meters, or 8,156 feet). This peak is another of Mount Mazama's cut-in-two satellite volcanoes. Below its cliff it displays a spired, structureless mass of volcanic rock that cooled in its central conduit.

Llao Rock. The great mass of Llao Rock and the wings of volcanic rock that spread to either side were once interpreted as a single, massive lava flow filling a deep, U-shaped glacial valley. Recent research, however, suggests that the rock is the lava-filled crater of a satellite volcano essentially contemporaneous with Mazama's climax eruption. The massive gray lava is sandwiched between an earlier pumice layer and light-colored pumice known to come from Mazama's final explosions.

Sun Notch (far right) and Kerr Notch bear in their U-shaped profiles the hallmark of glaciation. The beheaded valleys are clues to the former height of Mount Mazama, for the ice that shaped them must have accumulated at much higher elevations.

Recent research suggests that Llao Rock's vertical cliff displays a lava-filled crater almost contemporaneous with Mount Mazama's climax eruption.

Mount Scott (elevation 2,720 meters, or 8,926 feet). A stratovolcano some distance removed from the caldera rim, Mount Scott escaped being cut in half as Mount Mazama collapsed. Glacial shaping of its western face shows that there was a time when it supported small glaciers. From its summit can be seen the many cinder cones that dot the eastern and northern parts of the national park, and some dacite domes beyond the east park boundary. Dacite lava is usually viscous and pasty, and it squeezes out onto the surface like toothpaste out of a tube, forming a dome rather than a cone-shaped hill.

Phantom Ship. This unusual feature, one of Crater Lake's two islands, is thought to be part of the conduit system for Mount Mazama. Its rock, and the rock that it cuts, may be the oldest in the park. Boat-trippers may be able to discern the sloping flows of the ancient stratovolcano in the nearby crater wall; it was eventually covered over with layers of lava and pyroclastic debris of the growing Mount Mazama.

Pinnacles. Straight from the Land of Oz, these tapering spires are made of tuff from an avalanche of incandescent ash emitted during Mount Mazama's climax eruption. As the pumice cooled, steam and volcanic gases rose through it (as they have been observed to do near Mount St. Helens), cementing its particles together. Some pinnacles still

An avalanche of incandescent ash provided material for these tall pinnacles of pumice, ash, and fine rock debris. Hot steam filtering up through the volcanic ash strengthened it into chimneys that now resist erosion.

have thin, hollow channels up their centers. You can think of them as fossil gas vents or fumaroles.

The curious shape of the pinnacles, however, is not the only interesting feature of this ash-flow tuff. Of far greater importance geologically is the zoning displayed in the tuff. Light-colored, silica-rich rhyodacite ash at the base of the thick tuff layer grades smoothly upward into dacite, andesite, and then basaltic andesite, revealing that in one continuous eruption deeper and deeper levels of the magma chamber were being tapped.

Red Cone. A short distance northwest of Crater Lake, and visible from the Rim Drive, this cinder cone erupted shortly before Mazama's climax eruption. Because its cinders are basaltic, its magma is thought to have come from a separate, deeper source, one which it may have shared with Desert Cone, Bald Crater, Timber Crater, and other basalt cones in this part of the park.

Rim Drive. Good views of Crater Lake and its surroundings can be found along this highway. Displays at the viewpoints interpret many geologic features.

The Watchman. Drawing its name from its service as a survey post during the first sounding of Crater Lake, the Watchman offers some of the best views of Wizard Island. The peak itself, accessible

Large rockslides on the face of the Watchman shelve out at water's edge. The dacite lava flow that forms this peak can be traced downward to a feeder dike on the caldera wall.

Lava flows at Wizard Island's base developed ridgelike retaining walls as cooling lava piled up. Flow surfaces are marked by concentric ribs such as one might expect on a smaller scale in flowing cake frosting.

by trail, is, like many other high points along the rim, half of a satellite volcano. Its broken face is covered by wide, high rockslides. From the top, these slides look nearly vertical. Actually, they slope about 40 degrees, the angle of repose for the rocky fragments, still far too steep for safety. At water's edge they lose their steepness and shelve out horizontally toward Wizard Island. In winter and spring, tumbling rocks skid across deep snow and are deposited farther out in the lake, contributing to the shelf.

Wizard Island. Wizard Island formed during Mount Mazama's last (to date) gasp, probably no more than a thousand years ago. It is a typical cinder cone, even to the lava flows at its base, similar in form to its brothers in Lassen Volcanic National Park, Lava Beds National Monument, and other parks and monuments. Boat tours stop here. Some passengers elect to climb the cone and return on a later boat. The slopes are composed of cinders and volcanic breccia largely made up of bubbly dark lava called scoria. Some fragments show an abrupt transition between solid, bubble-free lava and scoria. Volcanic bombs thrown from the crater, some of them football-shaped from their spin through the air, can be recognized among the cinders. The bright red color of some of the cinders results from oxidation of iron minerals.

The crater, 100 meters (300 feet) across, has changed little since this small volcano's active days. From the summit you can get a good view of the lava flows that poured from the island's west base.

OTHER READING

Cranson, K. R. 1982. *Crater Lake—Gem of the Cascades.* K. R. Cranson Press, Lansing, Michigan.

Williams, Howel. 1941. *Calderas and their Origin.* University of California Department of Geological Sciences Bulletin, vol. 25.

Williams, Howel. 1942. *The Geology of Crater Lake National Park, Oregon.* Carnegie Institution of Washington, Publication 540.

Williams, Howel. 1954. *Crater Lake, the Story of its Origin.* University of California Press.

Williams, Howel. 1956. "Crater Lake." On reverse of topographic map, *Crater Lake National Park and Vicinity, Oregon.* U.S. Geological Survey.

Williams, Howel. 1962. *The Ancient Volcanoes of Oregon.* University of Oregon Press, Eugene, Oregon.

Devils Postpile National Monument

Established: 1911
Size: 3.24 square kilometers (1.25 square miles)
Elevation: 2300 meters (7600 feet)
Address: c/o Sequoia and Kings Canyon National Park, Three Rivers, California 93271

STAR FEATURES

• Part of an area where frequent earthquakes and recent volcanic activity result from continuing uplift of the steep eastern front of the Sierra Nevada.

• A superb example of columnar jointing, with slender polygonal columns up to 20 meters (60 feet) tall.

• Rainbow Falls, where the Middle Fork of the San Joaquin spills over resistant lava flows.

• Evidence of two or more periods of Pleistocene glaciation.

• Visitor center, trails, naturalist talks, and guided walks.

SETTING THE STAGE

Once thought to be a lava "waterfall" frozen in midair, Devils Postpile is now known to have formed by shrinkage of ponded lava. Its 20-meter (60-foot) columns are part of a lava flow that extends from Upper Soda Spring (outside the national monument) down the valley of the Middle Fork of the San Joaquin River to a point just south of the Postpile itself. The rock that makes up the flow

Columnar jointing of lava flows rarely reaches the perfection shown at Devils Postpile.

is dark gray basalt containing abundant white feldspar crystals. The basalt columns are thought to have developed where newly erupted, still fluid lava stood in a pool behind some sort of natural dam; there the lava cooled more slowly than in other, thinner parts of the flow. Cooling downward from its surface and upward from its base, lava develops vertical shrinkage cracks that eventually, growing upward and downward from these surfaces, give it a columnar or postlike appearance.

Ideally, shrinkage cracks in lava should form a perfect hexagonal pattern, the pattern that best relieves the tensions in shrinking material. More than half the columns here *are* six-sided. Others, however, are three-, four-, five-, or seven-sided, probably because irregularities in cooling upset the perfect pattern. Ideally, also, the columns should all be straight and vertical, because shrinkage cracks form at right angles to cooling surfaces. Curved columns, prominent at the north end of the Postpile, were caused by irregularities in the terrain over which the lava originally flowed.

Columns visible from the trail on the valley floor represent only the lowest part of the original ponded lava, the part that cooled from the bottom upward. The lake of lava is thought to have been 120 meters (400 feet) deep, and to have filled the valley from side to side; most of it has now eroded away.

The top of the Postpile—not in sight from the valley floor but accessible by a branch trail—has

Most of the columns are six-sided, though some have four, five, or seven sides.

Smoothed, striated, and polished by glaciers, the tops of the columns record an interesting interplay of fire and ice.

been smoothed and polished by a glacier that came down the valley long after the lava had solidified. Upper surfaces of the polygonal columns, exposed there in cross section, bear both glacial polish and glacial striae.

The black basalt of Devils Postpile was not the only lava to flow down this valley. A short distance downstream from Devils Postpile, and extending to Rainbow Falls and beyond, is a gray volcanic rock considerably lighter in color than the Postpile Basalt, that when weathered forms lots of small, irregular, more or less horizontal plates. At Rainbow Falls this platy rock is crudely columnar as well. The rock is dacite, intermediate in composition between rhyolite and basalt.

Another basalt flow makes up the Buttresses, west of the Middle Fork and opposite the dacite flow. As we shall see, it is the oldest volcanic rock in the area—a black basalt containing amber-colored crystals of olivine. All these volcanic rocks rest on granite of the Sierra core.

Mammoth Mountain, east of the national monument, represents a lava dome similar to the dome of Lassen Peak in Lassen Volcanic National Park, as well as to domes that developed in the crater of Mount St. Helens after its May 1980 eruption. Lava domes form from particularly thick, sticky lava, in this case dacite, that doesn't flow in the normal sense of the word, but mounds up over a volcanic vent like a giant ball of dough, its pasty, molten interior here and there breaking through its hardening crust.

Light gray pumice scattered on the ground both within and outside of the national monument is the product of explosive volcanism outside the monument area. This rock floats! And because it floats it can be carried quite far by streams or by sheetwash after severe rains.

At Rainbow Falls, 2.5 kilometers (1.5 miles) below Devils Postpile, the Middle Fork of the San Joaquin River leaps over a lava cliff into a large plunge pool 30 meters (100 feet) below. The falls have an interesting story, closely tied in with the volcanic and glacial history of the area. After the last glacier retreated, about 10,000 years ago, the Middle Fork of the San Joaquin River flowed down a canyon about 500 meters (1640 feet) west of its present channel, at the foot of the Buttresses. There, its canyon was bordered on the east by cliffs of platy andesite, and on the west by granite.

The river's course then shifted eastward, perhaps because of a landslide or some other obstruction, and it abandoned part of its little canyon. About 150 meters (500 feet) west of the present falls, it met its earlier channel once more, and took a flying leap down the lava cliff of its former canyon. The cascade itself—a prototype Rainbow Falls—constantly eroded away the loose, platy rock at the base of the cliff, undercutting it in such a way that stronger upper parts of the cliff repeatedly collapsed. Every collapse moved the falls upstream, until they arrived at their present position. In all likelihood they will continue to move upstream for some time to come.

The Middle Fork of the San Joaquin River surges over a rhyodacite cliff at Rainbow Falls. The rhyodacite is imperfectly cut with columnar joints.

Other interesting features in this area are several soda springs and hot springs—indications of a volcanically active area. Soda springs develop where magma cooling well below the surface gives off carbon dioxide—the same gas that makes ginger ale "fizzy." Hot springs derive their heat from hot rock, either directly as steam from the magma condenses, or indirectly as rainwater and snowmelt, percolating downward toward the magma, pick up heat that is passing upward from it.

GEOLOGIC HISTORY

Mesozoic Era. Though the geologic record of this region begins in Paleozoic time, with deposition of marine sedimentary rocks along the western edge of the supercontinent Pangaea, we will take it up in Jurassic and Cretaceous time, when the continent's westward drift brought about intrusion of the Sierra granite. Made up of many individual granite plutons—big balloon-like masses of molten rock that rose separately from the depths—the granite dates from about 200 million to 80 or 90 million years ago. The intrusions folded and altered overlying rock layers and then cooled slowly below the Earth's surface. Portions of the granite surrounded so-called roof pendants of metamorphosed older rock, which hung down between individual plutons like the pockets of air caught between bunched balloons.

A long period of erosion followed, during which most of the altered Paleozoic rocks of the pre- or proto-Sierra were eroded away. A few roof pendants remained; the dark, ragged crest of the Ritter Range, visible from the Mammoth Pass road into Devils Postpile, is one of them.

Cenozoic Era. Not until the end of the Cenozoic era, less than 5 million years ago, did the present Sierra Nevada begin to form. Then, the great block of the Sierra, edged by north-south fault zones, rose slowly, in many increments, to its present elevation. Because uplift along its eastern margin was much more rapid than that along its western edge, the east slope is the steepest, and the summit of the range lies well to the east. (For a time, it lay even farther east: Some of the present stream valleys, the Middle Fork of the San Joaquin among them, have been "beheaded" by fault movement and rapid erosion along the eastern face.) As the rock rose, continuing erosion stripped more rocks from its summit, and carved deep valleys into its core.

Along fault zones that edge the great Sierra block, particularly along those that define its eastern face, magma rose to the surface. In Pleistocene time, volcanism combined with glaciation, in a pattern well displayed in Devils Postpile National Monument. The oldest lava flow in the monument area is the basalt of the Buttresses, along the monument's western boundary. This basalt, which flowed out onto a glaciated granite surface, came from a source somewhere to the northwest. The Rainbow Falls Dacite, somewhat younger, also flowed onto a glaciated surface.

Another period of glaciation followed, with flowing ice smoothing and polishing some of the lava surfaces. Then, about 700,000 years ago, more volcanism: Explosive eruptions near Reds Meadow created the Reds Meadow Tuff. And more glaciation, followed by intrusion of the stiff, viscous dacite lava dome of Mammoth Mountain.

About 100,000 years ago, the Devils Postpile Basalt erupted from a vent near Upper Soda Springs, north of the national monument. The lava, flowing down the Middle Fork Valley for about 5 kilometers (3 miles), was dammed, as we have seen, by some natural obstacle, so that a deep pool formed, setting the stage for development of the lava columns.

Sometime after the Devils Postpile Basalt had hardened completely, another glacier, its source high in Sierra peaks, flowed down the Middle Fork, rounding, smoothing, and polishing exposed surfaces of the lava flow, as well as deepening and broadening its route by freezing to and then pulling away many of the lava columns.

About 10,000 years ago, melting of the ice signaled the end of Pleistocene time. Not long after that, eruption of cinder and lava from several small volcanic vents southeast of the national monument established the Red Cones and their associated lava flows. And even more recently, perhaps with the explosive volcanism that led to development of nearby Mono Craters, light gray pumice—hardened volcanic foam—was scattered over the region, to crunch underfoot as you walk the trails of Devils Postpile National Monument.

OTHER READING

Eckhardt, Wymond W., and Huber, N. K. 1985. *Devils Postpile Story.* Sequoia Natural History Association.

Huber, N. K. and Rinehart, C. D. 1965. *Geologic Map of the Devil's Postpile Quadrangle, Sierra Nevada, California.* U.S. Geological Survey Geologic Quadrangle Map GQ-437.

John Day Fossil Beds National Monument

Established: 1974
Size: 60 square kilometers (23 square miles)
Elevation: 616 meters (2,019 feet) at Cant Ranch Visitor Center
Address: 420 W. Main, John Day, Oregon 97845

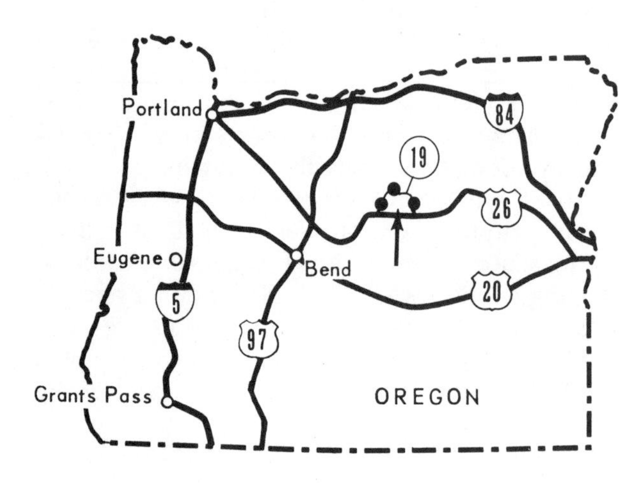

STAR FEATURES

• Gaudy layers of Cenozoic volcanic rock, including lava flows and fossil-bearing volcanic ash. Fossil plants and animal remains found here show adaptations in plant and animal life that accompanied profound climatic changes 50 to 30 million years ago.

• Evidence of sporadic but long-lasting volcanism, with ash falls and mudflows similar to those resulting from the Mount St. Helens eruptions of 1980.

• A look at a hybrid geologic area with faulted mountains resembling those of the Basin and Range region to the south, and horizontal lava flows of the Columbia Plateau to the north.

• Badland erosion, with examples of differential weathering of hard and soft rocks, and with landslides and rock falls.

• Excellent wayside and visitor center exhibits, guided walks, and self-guided nature trails clarifying geologic features.

SETTING THE STAGE

North of the folded and faulted mountains of the Basin and Range Province, south of horizontal basalt flows of the Columbia Plateau, lies a region

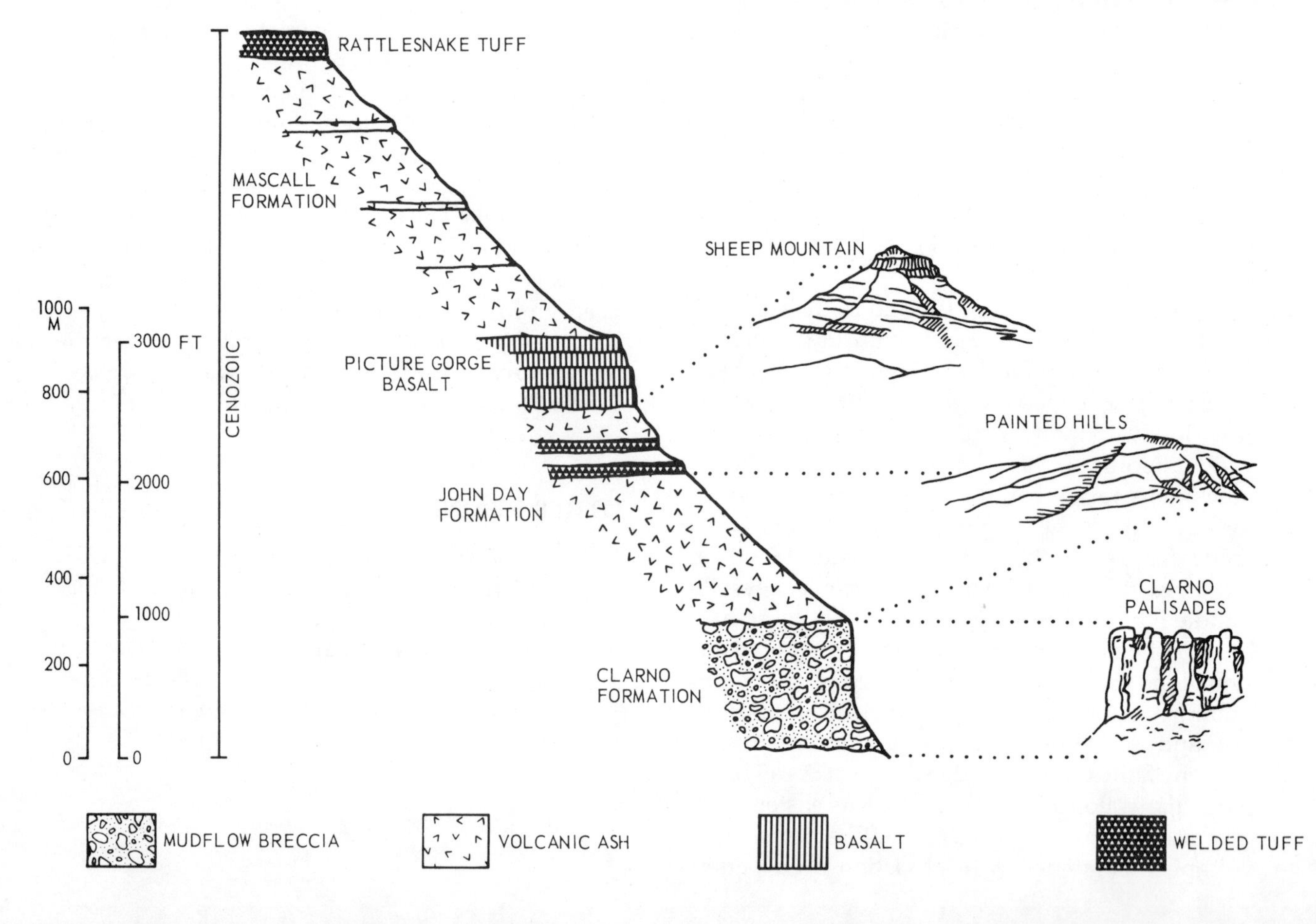

Palisades of the Clarno Unit weather from a single mudflow-mudslide deposit, an unsorted mass of rock fragments and volcanic ash. The talus slope below the cliff is a good place to find and photograph fossil leaves and petrified wood.

which shows features of both: tilted fault blocks superimposed by somber lava flows. Colorful faulted wedges of soft volcanic ash are exposed in valleys and bluffs of the John Day River and its tributaries. These rocks bear in remarkable sequence many relics of the past, evidence of widespread and often violent outpourings of volcanic ash, of destructive mudflows, and of exotic plants and animals struggling to survive amid ash falls, lava flows, and unpredictable changes in climate.

Each of the three units of John Day Fossil Beds National Monument tells its own part of the geologic story. The Clarno Unit's rugged brown bluffs and pinnacles of coarse rock formed early in Cenozoic time from a mudflow-mudslide deposit. The volcanic cobbles on the bluffs, rounded by tumbling water, were carried in a dense slurry of mud and sand like cement fresh from the mixer. Thin beds of fine volcanic ash, probably deposited in ponds dammed by the mudslide, contain leaves and branches of trees destroyed in the volcanic cataclysm.

The Painted Hills Unit displays somewhat younger layers of ash, the brightly colored John Day Formation. In it, too, fossils have been found. But in the soft ash they are hard to see. Geologists search gullies for tiny bone fragments and then work up the gullies to find their source. The self-guided trail here points out geologic features.

The John Day Formation is particularly well exposed in the Sheep Rock Unit. Rust-red, brown, buff, and green ash beds, the accumulation of many millenia, are topped by resistant lava flows of the Picture Gorge Basalt. These flows are part of the vast Columbia Plateau, one of the largest volcanic expanses in the world. Above the basalt is the Mascall Formation, white, water-deposited volcanic ash.

This national monument was established to protect world-famous animal and plant fossils found in the volcanic ash beds. Plant fossils are in the form of petrified logs, leaf and stem impressions, and seeds and nuts. Animal fossils, a good deal rarer, occur as bones and teeth. All are protected by national monument status, and unauthorized collecting is not allowed.

Many factors contribute to the abundance of fossils here. At a time when mammals were in the

Colored beds of the John Day Formation yield plant and animal fossils that help geologists piece together the scenery and climate of bygone days.

Picture Gorge Basalt caps Sheep Rock. The lower dark layer is welded tuff. Note the diagonal fault (arrow) that offsets the strata here.

ascendancy, this region was as thickly forested as the subtropical rain forests of today. Volcanic explosions—some of them many times greater than the Mount St. Helens eruption of 1980—spread enormous volumes of volcanic ash across the land. Much of the ash fell directly to the surface; some was carried by churning slurries of ash and water and rock that swept down stream valleys (again with parallels at Mount St. Helens). Leaves caught up by the mud flows were mashed together and rolled up like legendary bankrolls. Fleeing animals fell victim to hot gases, ash, and roiling mud. Others mistook the abnormal darkness of ash clouds for night, and were entombed as they slept. Downed trees, thoroughly battered, were buried by the floods.

Rock units that make up the Tertiary sequence here were deposited in a low area between the Cascade volcanoes and the Blue Mountains. Geologists now call this area the John Day Basin, honoring its past basin shape rather than its present mountainous one. The basin was later destroyed by crustal movement that broke and tilted the once-flat ash layers, and by lava flows that covered the broken hills with thick, protective sheets of basalt. Subsequent erosion eventually cut through the protective layer to create the landscape you see today.

Weathering processes operate at vastly different rates in hard basalt flows and in poorly consolidated layers of volcanic ash. It is this difference that regulates the present landscape. Volcanic ash decomposes into bentonite, a type of clay that swells with every rain and shrinks as it dries. The fine, powdery surface is continually loosened, to be whirled away by wind or washed downslope by the next rain. Ash still glowing as it fell fused into a harder rock type called welded tuff, a material that does not alter to bentonite but stands up as bluffs, ledges, and sometimes palisades. Basalt lava flows and dikes, which are still harder, resist erosion well. The narrow gorge of the John Day River as it cuts through Picture Gorge Basalt, in the Sheep Rock Unit of the monument, contrasts markedly with the broad valleys and low-profile hills shaped by the same river in softer formations. Weak rocks often skid downslope here, carrying part of the hard cap-rock with them, as at Cathedral Rock. Landslides help to widen today's valleys.

Many ash layers are brightly colored by small amounts of iron minerals disseminated through them. Some of these iron minerals were part of the original ash; others were introduced later by groundwater seeping slowly through the rock. The red mineral is hematite, an oxide of iron. Yellow and buff colors are due to limonite, a related mineral group in which each iron oxide molecule is closely associated with a water molecule. Green iron compounds tend to form after burial of rocks that contain plentiful organic (animal and plant) material, as in layers deposited in marshes and swamps. Thus even the color of these ash beds gives clues to the environment in which they accumulated.

Clay minerals in altered volcanic ash swell when wet, shrink when dry. This is the result. Few plants can gain a foothold on the cracked, crusty surface. (Dime shows scale.)

GEOLOGIC HISTORY

Cenozoic Era. Almost all the rocks in the national monument are Cenozoic, and almost all are volcanic. All are terrestrial or continental, deposited on land rather than in the sea. Farther east, the long ranks of the Rocky Mountains rose early in Cenozoic time, in their successful bid to become the backbone of the continent. Westward the land sloped gently to the sea. The whole continent was drifting west as the Atlantic widened. Periodically on its western slope volcanoes flung ash high into the sky or shot out incandescent avalanches of ash mixed with buoyant, rapidly expanding gases. Such explosions laid waste the countryside, flattening forests and suffocating plants and animals. Clouds drawn together by convection above the volcanoes added deluges of rain, and great mudflows rafted chunks of volcanic rock through lowland valleys, leaving desolation in their wake. Though forests died, trunks and leaves swept by the torrents survived as fossils.

The earliest eruptions took place about 54 million years ago. Dense subtropical rain forests, with large-leaved, thick-leaved trees related to avocado, pepperwood, cinnamon, fig, and sycamore, fell before them. Unusual metasequoias, needle-shedding relatives of our giant redwoods, fared no better. We can deduce from modern cousins of these trees that central Oregon at that time was quite unlike central Oregon today. Such trees need plenty of rain, mild temperatures, and absence of frost. In lush forest glades roamed strange animals only distantly related to modern forms: bulky titanotheres, predatory creodonts, tiny, delicate, four-toed horses, and long-snouted tapirs.

The volcanic ash layers and mudflow deposits containing these fossils make up the Clarno Formation, 54 to 36 million years old. These rocks were folded, faulted, and deeply eroded before the next volcanic episode began. By about 36 million years ago the climate was slightly cooler, but there were still no Cascade or Coast Ranges to waylay moist winds from the Pacific. Oak, chestnut, birch, and beech grew in a forest dominated by the metasequoia. Between 36 and 20 million years ago, new volcanic centers showered the landscape again and again with ash, ultimately with the hundreds of meters of it that are now the John Day Formation. Leaves, petrified wood, and bones from this formation tell us of still lush vegetation and still numerous animals. Herds of oreodonts, sheeplike creatures that left no modern descendants, browsed the shrubbery. Giant pigs eyed lumbering rhinoceroses while tiny rodents and three-toed horses slightly larger than their four-toed predecessors scurried among the roots.

By the middle of Cenozoic time the coastal ranges had begun to rise. Westward movement of the continent was carrying it over the East Pacific Plate, scooping up sea-bottom sediments that piled in crumpled masses along the continent's leading edge. Eventually, in the prolonged collision, mountainous masses of sea-floor scrapings rose high enough to cut the interior off from Pacific moisture. Vegetation of Cenozoic time, as well as its animal life, reflects the increasing aridity. Grasslands spread, tempting forest denizens with new foods and luring them away from protective shadows into the open, where only swift flight could save them

As horses moved out onto grasslands, they became larger and longer legged, and their teeth became especially suited to a harsh diet of prairie grasses.

On Crater Lake's wall, lava flows alternate with thin layers of volcanic ash. Red edging on the lava flow is soil fired like pottery by the lava's heat.

Crater Lake's waters are nearly 600 meters (2,000 feet) deep. Their remarkable clarity derives in part from year-round coldness, which discourages most forms of microscopic life. Note the two U-shaped notches in the rim; they tell a story of a taller mountain, draped in ice.

The oldest rocks visible in the caldera walls are at water level near the Phantom Ship. The ship itself may have been the conduit of the early volcano.

Lassen Volcanic National Park

Fumes from underlying lava tinted Painted Dunes. Discolored areas are sites where short-lived fumaroles have oxidized minerals in the cinders.

Rocks of Chaos Jumbles fell from Chaos Crags, in the background. Riding a cushion of compressed air, they avalanched downslope, possibly at speeds of 150 kilometers (100 miles) per hour.

Brokeoff Mountain in the distance was part of ancient Mount Tehama. Pink dacite in the foreground fell from Lassen Peak dome.

Mount Rainier National Park

Mount Rainier's gleaming crest spawns the very glaciers that are destroying it. For more than 100 years it has lain dormant, emitting only a little heat and steam.

The green color of the Ohanapecosh Formation makes it easy to recognize. Small dark fragments are bits of fine volcanic mud caught up in undersea mudflows.

Longmire's hot mineral springs are tinted with algae able to survive in the hot water.

Snowfields of the Tatoosh Range, brown with Mount St. Helens ash when this picture was taken, harbor small ice fields that may once have been glaciers.

Mount St. Helens National Volcanic Monument

The line between killed and living forest is sharp and clear. Still-green parts of the forest were sheltered from the lateral blast by an intervening ridge.

Now horseshoe-shaped, Mount St. Helens looks down on barren, pumice-covered, once forested ridges.

Golfball-sized pellets of gray pumice reached ridges several kilometers from the eruption site. By 1985, new life is returning to the barren land.

Sea stacks along the Pacific coast retain a flat upper surface that is part of a long-ago wave-cut terrace now lifted above the sea.

Beach pebbles and "ruby" sand—actually tiny grains of pink garnet—tell of glacial boulders tumbled to the coast by ice-fed torrents draining the central ranges of the Olympic Peninsula.

As the coast recedes before the endless surf, trees lean and topple to join wave-tossed comrades on the beach.

Redwood National Park

South of the Klamath River, where river sand is distributed by a southward current, beaches are common.

Cliffs, sea caves, and pebbly pocket beaches edge the coastline north of the Klamath River's mouth.

The Gold Bluffs, golden in color when compared with the dark Franciscan rocks elsewhere along the coast, also contain gold.

The Marble Fork of the Kaweah bubbles among stream-worn boulders. Marble ledges farther downstream give the river its name.

Giant Sequoias are "living fossils" whose Tertiary ancestors were abundant worldwide.

Jointed granite decomposes underground to grus, a coarse quartz-and-feldspar sand. Blocks of granite are rounded in the process.

Yosemite National Park

The waters of Hetch Hetchy Reservoir now hide the valley of the Tuolumne River, once a second Yosemite. Tueeulala and Wapona Falls leap from hanging valleys.

Yosemite Falls vault from a hanging valley, the site of a tributary glacier in El Portal time.

Initially shaped by exfoliation, Half Dome was sliced in two by Tenaya Valley glaciers.

from predators. Three-toed horses, with their foot bones lengthened for added speed, had moved out onto the grasslands.

About 17 million years ago, volcanism took a new turn. Dark, fluid lava welled from narrow fissures to completely flood the land. These were the Columbia basalts of Washington, Oregon, and Idaho—flow upon flow, widening into lakes of molten rock, cooling into the vast Columbia Plateau.

As volcanism once more abated, streams and ponds left beds of ash atop the great plateau. These beds, too, yield plant and animal fossils, and they tell of a yet cooler, yet drier climate 15 million years ago. Deciduous trees—hickory, maple, oak, and elm—shed their leaves to survive the frosts of autumn and winter. Grasses claimed new territory. Horses grew to pony size and became ever more efficient runners, with teeth adapted to eating the harsh prairie grasses.

About 6 million years ago, storms of incandescent ash sheeted across this part of the volcanic plateau, cooling as the welded tuff of the Rattlesnake Formation. This rock unit preserves fossil mammals closely related to ones now living: one-toed horses (which were soon to become extinct in America), camels (we guess from absence of desert deposits that they had no humps), rhinoceroses, peccaries, bears, cats, dogs, rabbits and small rodents.

Uplift then of much of the western part of the continent initiated a cycle of erosion that endures to this day. Except where they were protected by hard basalt layers, earlier rock units were stripped away and carried by deepening rivers to the sea.

The last event in this history began about 2 million years ago—the coming of the Ice Ages in Pleistocene time. Though the lowlands here were never glaciated, they were marked by great increases in precipitation that went with glaciation. Heavy runoff increased the cutting power of rivers, so that in places they broke through the protective basalt layers to undermine the ash below. Rains lubricated landslides, such as that at Cathedral Rock, and increased the tempo of erosion in other ways, too. Then, as recently as 10,000 years ago, precipitation decreased, temperatures rose, glaciers vanished from the highlands, and the climate and scenery became as they are today.

OTHER READING

Anonymous. 1976. *The Geologic Setting of the John Day Country.* U.S. Government Printing Office, Washington, D.C.

Chaney, Ralph W. 1958. *The Ancient Forests of Oregon.* University of Oregon Press, Eugene, Oregon.

Lassen Volcanic National Park

Established: 1907 as a national monument, 1916 as a national park
Size: 430 square kilometers (166 square miles)
Elevation: 1,597 to 3,187 meters (5,240 to 10,457 feet)
Address: Mineral, California 96063

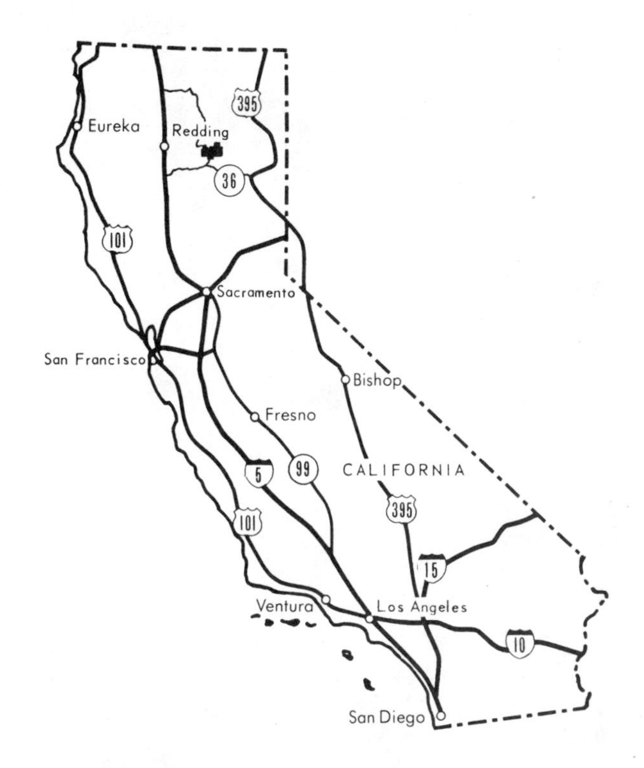

STAR FEATURES

- One of the world's largest lava domes, a mass of pasty lava squeezed through the eroded flank of an older volcano.
- Three other types of volcanoes. Virtually all the rocks in this park are volcanic.
- Vivid evidence of a violent eruption in 1915, when a destructive mudflow and a devastating "great hot blast" scarred the mountainside.
- Hot springs and mudpots still drawing heat from subterranean sources.
- An interpretive program that includes excellent wayside displays, a road guide keyed to numbered signs, guided and self-guided walks, evening programs. A new visitor center is planned.

See color section for additional photographs.

SETTING THE STAGE

Southernmost of the Cascade volcanoes, Lassen Peak is not as high in elevation as many of its cousins to the north. It differs from most of them in

Lassen Peak looks down on the area devastated by its 1915 eruptions. This photograph was taken in 1952. Forest growth is denser now.

Multiple craters at the summit of Cinder Cone indicate multiple eruptions. Cinder cone eruptions eventually peter out, usually after lava flows emerge from the base.

that it is a lava dome rather than a stratovolcano. Domes develop as thick, sticky magma pushes up through a conduit, bulges out at the top, and hardens in place. Lassen's magma, too thick and pasty to flow in the usual sense of the word, mounded into an unusually large mass that forms all the steep central portion of the mountain. Its grooved, nearly vertical cliffs may be the original surfaces of the hot magma. At the time of its growth, fragments of cooling lava broke from its swelling flanks to initiate growth of the talus slopes that now surround and partly obscure the cliffs. Unlike most domes, Lassen Peak has a conduit reaching up through its heart to a summit crater.

Lassen's dome rests on the remains of an earlier stratovolcano now called Mount Tehama, which in turn rests on a rolling plateau of volcanic rock dissected by stream valleys. Several jutting peaks southwest of Lassen—Brokeoff Mountain, Mount Diller, Pilot Pinnacle, Diamond Peak, and Mount Conard—are the remains of Mount Tehama.

The origin of the Cascade volcanoes is discussed more fully under Mount Rainier and Crater Lake National Parks. Suffice it to say here that the lavas of the Cascade volcanoes are chemically similar to continental rocks like granite and sandstone. They almost certainly originated where part of the continent was drawn down along a subduction zone to depths where remelting took place. Once melted, they worked their way to the surface again, for magma derived from the continents seeks to "float" on dark, heavy, iron-rich magma generated from oceanic plates.

Lassen's predecessor Mount Tehama was built of alternating layers of light gray lava and volcanic ash. Cone-shaped, it probably looked a great deal like Mount Hood. It was higher than Lassen by at least 300 meters (1,000 feet).

Even aside from Lassen Peak and Mount Tehama, this national park is a veritable outdoor classroom of volcano geology. Let's look for a moment at the origins of four types of volcanoes that occur here.

• Stratovolcanoes, built of layers of moderately fluid lava, breccia (broken volcanic rock), and volcanic ash, develop slowly because their activity is sporadic. As we've learned from Mount St. Helens, their eruptions can be both constructive and destructive, with setbacks in growth resulting from violent explosions. Some stratovolcanoes are almost perfect cones, with summit craters and steeply sloping sides. Others are surrounded by smaller parasite volcanoes that tap the same magma reservoir. The lavas usually are dacite or andesite. There are literally hundreds of stratovolcanoes around the world, the majority lying in the "Ring of Fire" that encircles the Pacific Ocean.

• Domes are common also; Lassen is one of the largest. Reading Peak, Chaos Crags, and several other mountains in the park are also lava domes. Their thick lavas thrust up like rising bread dough, disrupting surrounding rocks but not flowing far out over them. Domes form rapidly, all at once on a geologic time scale—sometimes within just a few days or weeks. Some permanently plug a volcano's vent, preventing further activity. Others live more dangerously, plugging the vent only until built-up pressures blow them to bits in the first moments of a new volcanic explosion.

• In shield volcanoes, successive flows of much more fluid lava (usually basalt) pour out in flowing rivers and harden in gently sloping layers. There's not much explosive activity, and volcanic ash is never abundant, though cinder cones may develop at the summit or on the flanks. Prospect Mountain and several other low-profile cones in this park are shield volcanoes.

• As for the fourth type, cinder cones, there is a perfect example in the northeast part of the park—Cinder Cone. This type of volcano rarely grows higher than about 300 meters (1,000 feet). As their name reveals, cinder cones are built almost entirely of volcanic cinders, popcorn-sized particles of bubbly scoria. In composition, the cinders are usually basalt; they may be black, dark gray, purple, red, even yellow, depending on the degree to which their iron minerals are oxidized. Scoria forms from the froth or foam at the top of the magma kettle. When pressure on the magma is released at the beginning of an eruption, volcanic gases dissolved in the magma come out of solution (just as carbon dioxide does when a bottle of carbonated beverage is opened) and form a light, bubbly, mildly explosive mass that tosses fine fragments of itself into the air. Falling to earth, these fragments pile up around the volcanic vent. If explosions become more violent, larger blobs of lava, as well as lava-coated chunks of rock broken from the conduit walls, spin through the air as volcanic bombs. Once the frothy upper portion of the magma is exhausted, fluid lava may rise in the conduit and force a passage through the side or base of the cinder cone, to flow out over the surrounding landscape. Both bombs and lava flows, stark and new, can be seen at Cinder Cone.

Lassen Peak itself bears little evidence of glaciation. However, glacial ice left its imprint on many of the older volcanic rocks exposed in and around the

Breadcrust bombs of many sizes litter the cindered ground near Cinder Cone's steep flank.

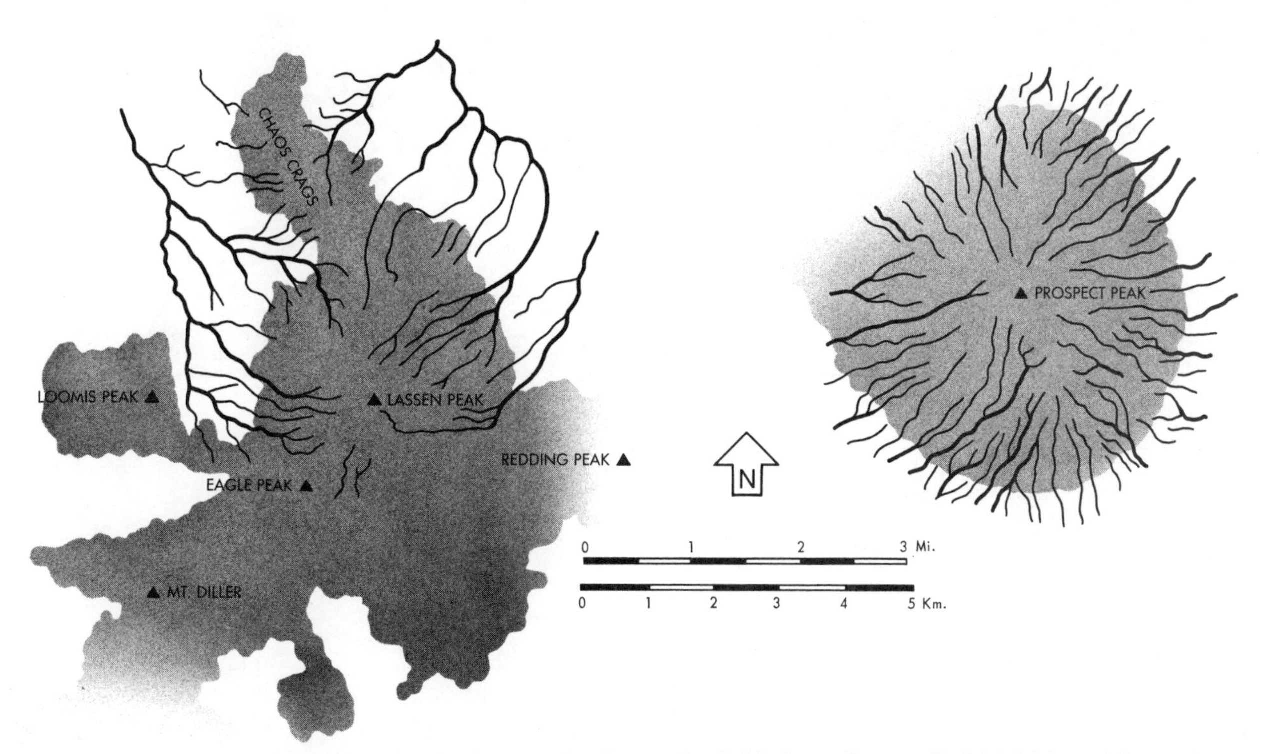

Volcanoes can often be recognized on maps by their circular shape and radial drainage. Prospect Peak (right) is a good example. Lassen Peak's drainage, however, is deflected northward by nearby peaks and ridges (left).

park: the modified U-shaped valleys of Warner and Hot Springs Creeks, glacial scratches and grooves on exposed ridges, lake basins scoured in rock, moraines, and scattered boulders moved by ice.

GEOLOGIC HISTORY

Cenozoic Era. Since no Paleozoic or Mesozoic rocks are exposed in this national park, let's start our discussion of geologic history at the beginning of the Cenozoic Era, about 65 million years ago. Taken as a whole, northwestern United States was rising at that time, and seas had drained away. The climate was warm and humid; trees of temperate and tropical varieties forested present-day Washington, Oregon, and northern California.

As the land rose, floods of basalt streamed over it, flow upon flow, building the vast Columbia Plateau of Washington and Oregon. Some of the basalts reached into northern California.

Then, 15 million years ago, volcanism took a new form. Less fluid, lighter-colored andesite lava began to erupt here and there through the dark basalt. The new lava tended to pile up to form discrete volcanic peaks whose layers of lava alternated with layers of volcanic breccia and ash.

Rocks within Lassen Park range from the old flood basalts to ash from 1917 eruptions. Those younger than 15 million years developed during six stages of volcanic activity:

1. Early eruptions built up a lava plateau now visible as a nearly level surface in central parts of the park.

2. In new outbursts of volcanic activity, thicker lava built several small, steep-sloped stratovolcanoes on this plateau, Hat Mountain, Crater Butte, and Fairfield Peak among them. At about the same time, the shield volcanoes of Prospect Peak, Raker Peak, and Mount Harkness developed.

3. Over thousands (perhaps millions) of years, the great stratovolcano Mount Tehama rose farther west, its central vent probably above today's Sulphur Works. A number of small outlying cones are thought to have tapped the same magma reservoir.

4. Thick, pasty dacite forced its way up into the andesite cone of Raker Peak, and dacite lava flowed from a new vent on Mount Tehama's northeast flank, cooling into glassy rock now called pre-Lassen dacite.

5. About 11,000 years ago, a single very large mass of sticky, pasty lava oozed up through the same or another vent. In a period of a few weeks or months, it grew to be the huge dome of Lassen Peak—the entire upper portion of the mountain. Nearby, several small domes appeared: Bumpass and Helen Mountains, Vulcans Castle, Reading and Eagle Peaks.

6. Within the last 1,000 or 1,200 years, renewed activity created the three domes of Chaos Crags, Cinder Cone erupted in the northeast corner of the

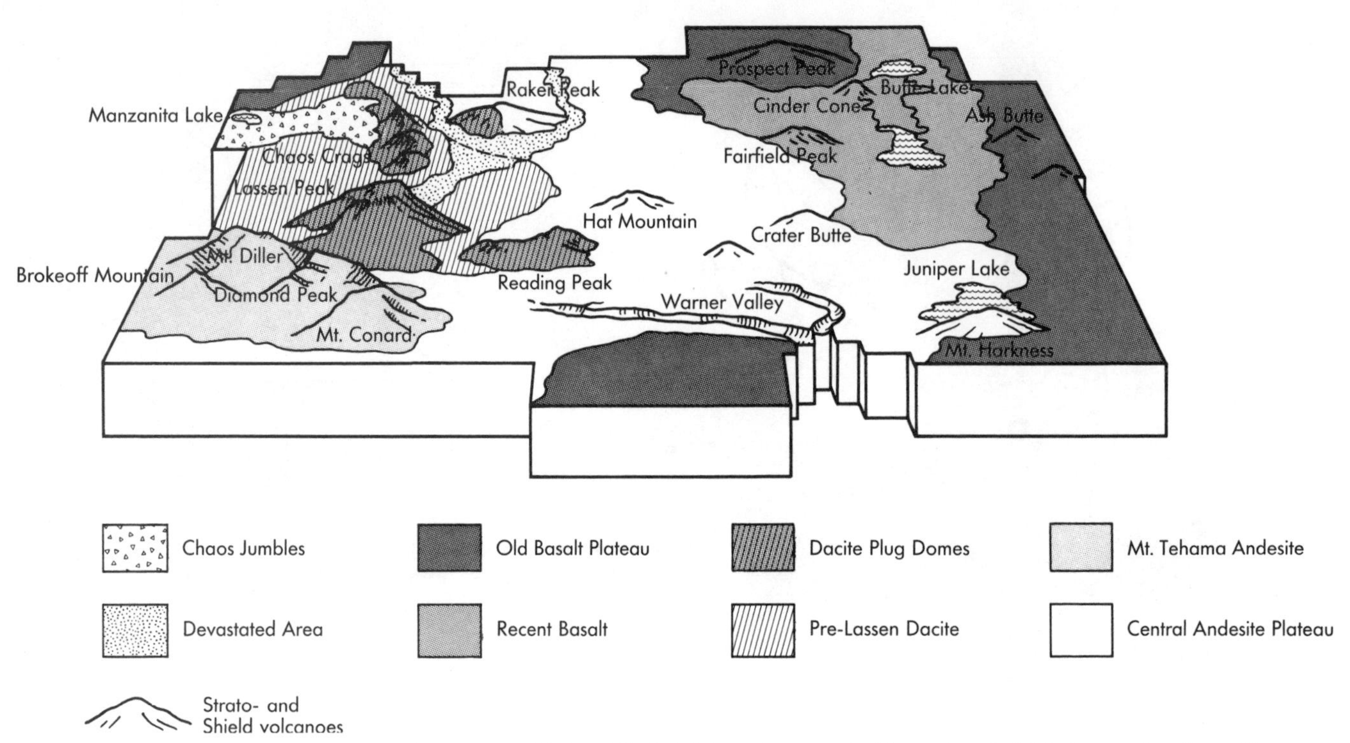

In simplified form the volcanic rocks of Lassen National Park fall into six recognizable units, of which five are less than 15 million years old.

park, and activity at the summit of Lassen Peak created a crater there. Prior to 1914, the crater was fairly large and deep. In 1914 this crater—a volcano on a volcano—came to life with a sequence of small steam eruptions, lava flows that filled the crater to overflowing, and explosions of hot ash and gas.

There still remains a big question: What happened to Mount Tehama? For a time, geologists assumed that it had destroyed itself in an explosion-and-collapse sequence comparable to that of Crater Lake's caldera in Oregon. But there is no widely distributed layer of Mount Tehama ash, as there is for Crater Lake's progenitor. Geologists now think it more likely that Mount Tehama was destroyed by erosion, the work of both streams and Pleistocene glaciers. If we reconstruct its probable height by projecting the slope of lava flows on Brokeoff Mountain, Mount Diller, Pilot Pinnacle, and Mount Conard upward until they meet, we come out with a volcano at least 3,300 meters (11,000 feet) high, quite high enough to support glaciers during the cool, wet cycles of the Ice Ages. Corroded by hot underground water, percolating steam, and volcanic acids, its rocks crumbled easily before the onslaught of the glaciers. A cirque below Brokeoff Mountain, once thought to have been the steep edge of the defunct caldera, confirms the former presence of glacial ice, as do the many glacial lakes, moraines, and striated rocks elsewhere in the park. Lassen itself, though, developed so recently that it was scarcely touched by Ice Age glaciation.

The other big question is, of course: Will Lassen erupt again? The question can't be answered, but geologists now feel (with studies in Japan, USSR, Hawaii, and most of all at Mount St. Helens to back them) that careful monitoring of earthquake activity can give ample warning of an impending eruption, probably many days ahead of time. Seismic monitoring is now routine at this and other Cascade volcanoes, so consider yourself safe unless warned to evacuate.

BEHIND THE SCENES

Brokeoff Mountain (elevation 2,814 meters, or 9,235 feet). The crags of Brokeoff Mountain display the sloping lava, breccia, and ash flows of Mount Tehama's western flank. The trail to this peak leads past lovely Forest Lake, a tiny glacial tarn. Most of the route is thickly forested, but views from the summit are splendid. Brokeoff Mountain, Mount Diller, and Vulcans Castle define the western side of Mount Tehama; Diamond Peak and Mount Conard are remnants of its eastern slope.

Glacial striae mark some of the park's older volcanic rocks.

Odorous steam rises from fumaroles, hot pools, and mudpots in the surrealistic landscape of Bumpass Hell.

Fantastic Lava Flow advanced slowly, spilling cooled and fractured fragments in front of itself in a caterpillar tractor sort of motion.

Bumpass Hell. This interesting hot springs area is not far from the highway, and a guide leaflet describes natural features along the trail.

Where subsurface water is heated by contact with hot volcanic rock or superheated steam and can escape rapidly to the surface (as in stream valleys or other depressions), it emerges in boiling pools, mudpots, and steaming fumaroles. In some hot springs, the hot water almost immediately finds a route downslope. In boiling pools, the hot water remains for a time in the pool, which usually overflows enough to flush away mud that would otherwise accumulate. Mudpots have little or no overflow; clay from decaying volcanic rock is not washed away. Most of the area around the springs is leached by hot water and steam carrying corrosive acids. Protect yourself and your camera lens by standing back from splashing pools, and by all means stay on the boardwalks provided for your safety. Very near the springs are deposits of siliceous sinter, a satiny variety of silica deposited where hot silica-bearing water cools and evaporates.

Chaos Crags and Chaos Jumbles. Perhaps a thousand years ago, after Lassen Peak had formed, three more domes rose along a line extending northward from Lassen. Their stiff, pasty lava cracked and fragmented as it moved upward, so that talus accumulated around the plugs as it had around Lassen.

The hummocky land northwest of Chaos Crags is known as Chaos Jumbles. Its rocky, hilly surface has no established pattern of drainage, which tells us that it is geologically very young. Possibly triggered by a steam explosion close under Chaos Crags, an avalanche of rock and sand shot out across gently sloping terrain, creating the hummocky surface. A trail leads to an apparent explosion crater at the foot of the crags, now occupied by a small lake.

The Jumbles formed only 300 years ago. They lie on older volcanic debris formed in other violent eruptions of ash and broken rock. On geologic advice, the Park Service closed the Manzanita Lake Visitor Center and lodge in 1974; new facilities will occupy a safer site.

Cinder Cone. A self-guiding trail from Butte Lake to Cinder Cone passes close to one of the lava flows of the Fantastic Lava Beds, which flow from an

Lassen's 1915 mudflow climaxed more than 170 eruptions in the previous year. It carried huge blocks of hot lava several miles from their source.

opening at the base of Cinder Cone. Formerly the area was occupied by a much larger lake. White deposits at the edge of the flow are lake-bottom ooze composed almost entirely of microscopic skeletons of tiny plants called diatoms.

Both cinder and lava are basalt, dull black or dark gray except where they are oxidized. They contain more quartz than the older basalt that underlies them. Small white spheres are bubble spaces, or vesicles, filled in with opal (mostly a nonprecious variety) and other minerals.

Climb Cinder Cone if you can. At the summit, walk around its double-edged crater, the result of two or more periods of activity. Look northwest to Prospect Peak, a shield volcano, and south and southeast to the Painted Dunes, which are not dunes at all but a hilly field of cinders that fell onto a hot lava flow. Volcanic fumes from the flow discolored the cinders.

The most recent eruption here was in historic time. In 1851 pioneer settlers reported "a great fire to the eastward of Lassen," probably the reflected light of a red-hot lava flow. From the summit you can clearly distinguish this flow, the black, dragon-shaped river that courses down the center of the older lava field. Descending the northeast side of Cinder Cone, you'll have a chance to look into the gloomy mouth of the dragon's lair—the vent of that flow. Lava is an excellent insulator, and the cave remains cool through all seasons of the year.

Breadcrust bombs lie scattered around the base of Cinder Cone. These lava-coated lumps of rock, thrown from the summit crater, bounced and rolled down its flanks after their brief flight. Their lava coating cracked as gases within them continued to expand, creating the crusty surface that gives these bombs their name.

Devastated Area. Climaxing the eruptions of Lassen Peak that began in 1914, some 1915 lava from Lassen's summit crater triggered a mudflow on the volcano's northeast flank. Mixtures of melting snow and ashy mud carried hot lava blocks down the slope in a roiling river, ripping out trees and scouring one flank of nearby Crescent Crater. Divided by the prow of Raker Peak, half the mudflow turned into Lost Creek, while the other half flowed up and across Emigrant Pass into Hat Creek. In

The broad swath cleared by the 1915 mudflow and great hot blast can be seen to best advantage from Lassen's summit. A wide river of pink rock and gravel still seems to flow down the mountainside.

Trees downed by the great hot blast are gradually rotting away, their wood contributing nutrients to the barren soil.

these two valleys the flow destroyed trees, fences, and several farm buildings—a familiar story if you read about the Toutle River mudflow near Mount St. Helens. Mudflows contain so much finely divided rock material that they are almost as thick as wet cement and can carry along huge boulders.

Three days after the mudflow, a new catastrophe struck. As a towering cloud of steam and ash boiled upward from the crater to drift eastward across Nevada, a searing blast of hot ash and volcanic gases shot down the slope already devastated by the mudflow. Wider than the mudflow, this great hot blast snapped six-foot tree trunks as if they were toothpicks and laid them in windrows along the flanks of Crescent Crater and Raker Peak. Heat from the blast melted what snow remained on the slope, sending more mudflows down the valley of Lost Creek.

Today, after 70 years, the Devastated Area is still distinguishable. Soil has developed in sheltered pockets. Young conifers rise among the decaying logs of the former forest. Deer find food here; badgers, chipmunks, and ground squirrels burrow in the stony ground beneath the rotting logs.

Lake Helen (elevation 2,488 meters, or 8,164 feet). Lake Helen lies in a glacier-scooped hollow surrounded by glacially smoothed, light gray andesite of Lassen's predecessor Mount Tehama. Many vertical joints slice the rock here. Water freezing in them has gradually pried the rock apart—an important first step in formation of soil.

The lake is sustained by snowmelt. Though it has no surface outlet, water escapes through the fractured rock around its basin.

Lassen Peak (elevation 3,187 meters, or 10,457 feet). Lassen is the easiest and safest of the Cascade volcanoes to climb—no cliffs, no glaciers, just an uphill walk on a good trail, with an altitude gain of about 600 meters (2,000 feet).

The trail goes past some of the grooved cliffs, partly shrouded in talus, that are the sides of Lassen's central plug. Perhaps you can imagine the great mass of thick lava pushing up and up, over a period of just a few weeks. Cracks in the cooling crust widen, and the bright red glow of molten rock is seen beneath. Steaming blocks of lava break from the cracking sides and vault in long arcs down the flanks of the growing mass. Constant tremors shake the ground, and the discordant thunder of moving magma and falling rock fills the air. Still the mass thrusts upward until, finally, it reaches its present elevation. Then the movement ceases; the mountain rests.

The highest point on Lassen Peak is on a remnant of the dacite plug that makes up Lassen's southeast ridge. From the summit you can look down on the pink scar of the 1915 mudflow and the great hot blast. At its lower end the scar is peppered with young trees, but upper portions near and above treeline show little change.

Lassen Peak's summit pinnacles, part of a lava dome, offer a 360-degree panorama of the park.

Before 1914 the crater was much deeper than it is at present. Eruptions that began that year filled it in and built two small new craters just north of the highest ridge. These were covered over with lava in 1915, the same lava that now juts up in rough, dark ridges. On May 19, 1915, this flow rose in the crater and spilled out at two summit notches. It flowed down the southwest side for about 300 meters (1,000 feet) below the rim; the dark lava patch can still be seen from Manzanita Lake. On the steeper northeast side, the lava broke apart and tumbled, still hot, onto thick snow that overlay loose layers of the previous year's ash, initiating the great mudflow of the Devastated Area.

The two craters northwest of the lava formed in 1915 and 1917. Steam still issues from small fumaroles within them, and pale yellow sulfur tints their walls.

At Lassen's summit, rough ridges of the 1915 lava are backed by a small crater formed later that year. The lava conceals two deep craters formed in 1914.

Manzanita Lake. Dammed by shattered rock that avalanched from Chaos Crags, Manzanita Lake is only a few hundred years old. Already it has a sizeable delta at its south (upstream) end. Eventually it will fill in completely with delta sediments and plant growth, if it is not obliterated by another explosion-triggered avalanche of rock debris.

Prospect Peak (elevation 2,541 meters, or 8,338 feet). A trail starting near Butte Lake Campground leads to a lookout atop this shield volcano. Composed of fairly fluid basalt emitted during the second eruptive phase, Prospect Peak has at its summit a small, circular cinder cone. Its slopes are heavily vegetated, but from the lookout you can see Lassen's northeast face and the Devastated Area. Notice also several other shield volcanoes and small stratovolcanoes on the flat-topped plateau that surrounds Prospect Peak.

Raker Peak. An andesite shield volcano coupled with a dome of dacite, Raker Peak has a well defined crater at its summit. The cliffs on the side near the road are part of the dome.

During Lassen Peak's 1915 eruptions, Raker Peak cleaved the great mudflow and the great hot blast, deflecting them down both Lost Creek (now followed by the highway) and Hat Creek (the route of the Emigrant Trail). Trees felled by the great hot blast can still be seen on the slope of Raker Peak, though new vegetation almost hides the decaying trunks. One stand of older trees, sheltered by Survivors Hill, survived the blast.

Sulphur Works. Like Bumpass Hell, Sulphur Works has boiling pools, steaming fumaroles, and bubbling mudpots. This area is thought to be part of the central conduit of ancient Mount Tehama. As such, it lies beneath the point at which sloping volcanic rocks of Mount Diller, Brokeoff Mountain, Pilot Pinnacle, and Diamond Peak would have converged as they extended upward in the original volcano. You can get a really good look at these rocks—layers of lava, breccia, and ash now altered by the action of hot water and volcanic fumes—on Diamond Peak opposite the highway between Sulphur Works and the south entrance station.

Warner Valley. This lovely valley, as well as both its forks, is glaciated, but because volcanic rocks leached by hot acids are weak, the valleys are not as distinctly U-shaped as, say, some in Yosemite or Sequoia National Parks, where the rock is older, stronger, and more enduring. Hot Springs Creek in the western fork is fed by cold streams and springs whose water is only slightly warmed by discharge from Boiling Springs Lake and Devils Kitchen. The spa at Drakesbad uses water from several small hot springs nearby.

Boiling Springs Lake (there is a trail leaflet) stays at 51°C (125°F), so it is not really boiling at all; the surging motion is caused by rising steam bubbles. The water's odd color derives from fine clay particles, product of hydrothermal (hot water) alteration of surrounding rock. Satiny material around the lake is siliceous sinter, a form of silica. Hot springs in silica-rich volcanic regions slowly build up deposits of sinter as silica dissolved by hot groundwater comes out of solution when the water cools.

Other trails from Drakesbad lead to Devils Kitchen, an area of violently steaming fumaroles and boiling mudpots, and to Flatiron Ridge, a remnant of the lava plateau of the eastern half of the park.

OTHER READING

Crandell, D. R. 1972. *Glaciation Near Lassen Peak, Northern California.* U.S. Geological Survey Professional Paper 800-C, pp. C179–C188.

Crandell, D. R., and others. 1974. "Chaos Crags Eruptions and Rockfall-avalanches, Lassen Volcanic National Park, California," *Journal of Research,* U.S. Geological Survey, vol. 2, no. 1, pp. 49–59.

Loomis, B. F. 1926 (reprinted in 1945). *Pictorial History of the Lassen Volcano.* Loomis Museum Association, Lassen Volcanic National Park, Mineral, California.

Loomis, B. F. 1966. *Eruptions of Lassen Peak.* Loomis Museum Association, Lassen Volcanic National Park, Mineral, California.

Schulz, Paul E. 1959. *Geology of Lassen's Landscape.* Loomis Museum Association, Lassen Volcanic National Park, Mineral, California.

Schulz, Paul E. 1968. *Road Guide to Lassen Volcanic National Park.* Loomis Museum Association, Lassen Volcanic National Park, Mineral, California.

Williams, Howel. 1932. *Geology of the Lassen Volcanic National Park, California.* University of California Publications in Geological Science, vol. 21, no. 8, pp. 198–385.

Lava Beds National Monument

Established: 1925
Size: 189 square kilometers (73 square miles)
Elevation: 1,455 meters (4,773 feet) at visitor center
Address: Box 867, Tule Lake, California 96134

STAR FEATURES

• Volcanic landforms of varying age, some of them scarcely changed since their lava congealed and their cinders cooled.

• More than 200 lava tubes, of which 21 are easy to explore without guides. For do-it-yourself visitors, information sheets and flashlights are available at the visitor center.

• An introductory film in a lava-tube theater, roadside displays, self-guide descriptions of the caves, daily tours and talks (in summer only). Near the visitor center an outdoor exhibit displays the principal types of volcanic rock found here.

SETTING THE STAGE

The Modoc Plateau, part of the volcanic plateaus of Washington and Oregon, carries the geology of those areas southward into California. Here, volcanism took place in fairly recent time, some of it no more than 500 years ago.

Most of the volcanic rocks in this national monument are basalt, easily recognized as the dullest (in lustre, not interest) and blackest of rocks. Basalt contains more iron and less silica than lighter-

Schonchin Butte, a cinder cone, is cored with a basalt plug. The fire lookout is on the plug, highest and firmest part of the butte.

colored volcanic rocks like rhyolite, dacite, and andesite. Molten basalt is characteristically much more fluid and less explosive than magmas of those lighter rocks. Ordinarily thought of as the material of the oceanic crust, erupted along mid-ocean ridges, it occurs here because North America's westward drift has carried the continent over part of the Pacific Ocean's floor. As basalt oceanic crust sinks below the continent, some of it remelts and plumes upward to erupt as lava flows, cinder cones, and other volcanic landforms.

Many surface features of this national monument can be seen in perspective from the Schonchin Butte parking area or, better yet, from the fire lookout at the top of the butte. The whole of the monument is covered with lava flows, some partly concealed with dull, dark cinders. The flows spread across the northeast slope of an unprepossessing shield volcano, Medicine Lake Volcano, that rises southwest of the national monument. They erupted along several faults that define the west edge of Tule Lake Valley, the downdropped block of land whose lowest part is occupied by Tule Lake. The vertical offsets of the faults are easily spotted: Look north from Schonchin Butte toward Gillems Bluff; the faults show up as scarps or steps in the horizontal plateaus west of the bluffs and in the bluffs themselves.

Schonchin Butte is a cinder cone, with slopes of crunchy basalt cinders. Its summit crater is now partly filled by a lava plug on which is built the fire lookout. Frozen rivers of lava emerge from a fissure in its eastern face and flow north to the old shore of Tule Lake, now represented by a wave-cut bench well above the present lake surface.

Many other cinder cones are in sight: Hippo and Bearpaw Buttes to the south, Caldwell and Crescent Buttes to the southeast, Three Sisters to the east, and Whitney to the west. In a rough way you can tell their relative ages by the amount of vegetation on them, taking into account the differences between their more heavily vegetated north slopes and their hot, dry south slopes.

In this national monument one really gets a feeling for the behavior of molten basalt. Walk out over some of the flows for a close look at a few dark, pristine lava surfaces. Basalt lava is quite fluid and often spreads out in flat pools that later harden into resistant horizontal layers. It erupts quietly, not at all like the violently explosive lavas of the Cascade volcanoes. Dissolved gases—potentially explosive if they were to accumulate—left little bubble holes throughout the basalt, round or elongate depending on movement of the lava just before it congealed. Lava that is *very* bubbly is called scoria. Some bubble holes are filled with white minerals deposited from groundwater moving very slowly through the basalt long after it hardened.

By and large, molten basalt hardens in three basic patterns, two of them first described in Hawaii and given native Hawaiian names:

• The flexible, ropy-looking lava called pahoehoe is created when white-hot, molten basalt continues to flow beneath a thin and to some degree flexible skin formed on its cooling surface. The continued flow, which can be quite rapid, stretches and folds and twists the skin to a remarkable degree, as you'll readily see as you explore the monument.

• The incredibly rough, jumbled lava called aa cools in a thick, sticky, tarlike mass broken as it hardens by further movement of the flow underneath. Aa's jagged, loosely piled fragments are almost impossible to walk across. As thickening aa lava moves (often at only a snail's pace), rough fragments constantly tumble from its leading edge; it rolls right on over them, at the same time bringing new fragments forward on its upper surface.

• Blocky lava is topped by angular blocks with smooth faces formed by breakage of congealed lava.

There are about 200 known lava tube caves in the national monument, and probably many more that are unknown because they have no natural entrances. Without exception these long, vaulted tubes came into being when fluid lava flowed out from under a partly solidified crust. They vary quite a bit in size, complexity, and length. Twenty-one of them have been made more accessible by construction of pathways, stairs, and handrails. Only one, Mushpot Cave near the visitor center, is electrically lighted. With flashlights and something to protect your head from chance bumps on the jagged lava, you can explore as many caves as you wish, including most of the unimproved ones if you register first at the visitor center.

Cave walls and ceilings drip with solidified lava, and horizontal shelves in some caves mark lava levels that for a time were stable. Look for other details, too: colored mineral deposits, drawn-out bubble holes, lava "icicles," and, in a few caves, Indian pictographs. Some caves have smooth floors of congealed lava; in others the floors are covered with jagged breakdown from ceilings and walls. All the breakdown seems to have fallen soon after the caves drained; none is known to have fallen in historic time. The substantial rock walls and arched ceilings of these caves are almost certainly more durable than walls and ceilings built by man.

Lava tubes are welcome havens in the hot days of midsummer. Since cold winter air is heavier than warm air, it tends to flow into the caves and be trapped there. Some of the caves, especially those with deep levels, have ice in them year-round. Basalt is a good insulator and helps to preserve the ice. Outside the caves, tip up a block of lava that has been lying in the sun (watch, though, for snakes, who like to keep cool) and feel the difference between its upper and lower surfaces.

Another landform characterizing basalt eruptions is a type of volcano, albeit small: the spatter cone. As their name implies, spatter cones form when blobs of fluid lava splash and spatter from small volcanic vents, gradually piling up irregular cones of hardened lava that look somewhat like the

One of the 300 known caves in Lava Beds National Monument shows clearly the typical arched roof formed of the cooling surface of a lava flow.

As molten rock drains from a lava tunnel, dripping "icicles" of hot lava solidify on walls and ceilings. Later, groundwater seeping through cracks in the tunnel roof deposits white limy coatings near the cracks.

dribble-castles that children build with wet sand. Several spatter cones, their once-deep vents now clogged, line up along a single fissure at Fleener Chimneys; others occur at Ross Chimneys. Related features are the spatter ramparts near Black Crater and at the Castles. The red color around some spatter vents is iron oxide formed when the vents still exuded hot volcanic gases and steam.

GEOLOGIC HISTORY

Cenozoic Era. Looking out over the landscape of Lava Beds National Monument, one can readily surmise that the volcanic rocks and landforms are all quite young geologically. Some even appear brand new, as if erupted only yesterday. Many basalt and cinder surfaces show little in the way of soil development, and few plants grow on their almost barren expanses. Some flows have no soil at all, though they are beginning to be marked by colored lichens, pioneers of the plant world.

The oldest rocks are those of Juniper Butte, which erupted in Miocene time, about 15 million years ago. Unlike other park volcanic features, Juniper Butte is made of andesite, a light-colored volcanic rock more closely related to the Cascade volcanoes than to the Modoc Plateau basalts.

The history of this region goes back, however, to long before the time of the Juniper Butte andesite. It begins in Cretaceous time, more than 60 million years ago, when an early version of the Klamath Mountains of northwestern California broke away, or was torn away, from the long range of the early Sierra Nevada, and shifted about 100 kilometers (60 miles) west. Behind it, the sea moved in, and marine sediments were deposited. With passing millenia, the sea became more and more isolated as the Coast Range cut it off from its Pacific parent. Then it filled in with gravel, sand, and silt eroded from the mountains that surrounded it.

East-west tension seems to have lasted long after the Klamaths and the Sierra were torn apart, and it probably continues right up to the present time. About 30 million years ago, tension stresses opened up some long north-south fissures that reached deep down through the continental crust to tap basalt magma underneath the continent. Flow after flow of basalt poured from the fissures and spread over the sediment-filled basin, eventually building up the Modoc Plateau. Similar basalt plateaus developed at about the same time in Oregon and Washington, and near the California-Oregon state line their lavas merged with those of the Modoc Plateau.

After the building of these plateaus, even after eruption of Juniper Butte in Miocene time, some interesting and unusual volcanic events occurred in the Petroglyph Section of the monument, northeast of the main unit. In Pleistocene time, as rain and snowfall increased and decreased with the coming and going of glaciers farther north, Tule Lake grew and shrank and grew again. At a time when it was much larger than it is now (its old wave-cut shoreline can be seen near Hospital Rock), rising basalt magma somehow encountered the waters of the lake. The results of the encounters were what you might expect. Violent steam explosions near the surface created a cluster of shallow maar volcanoes, flat-floored craters edged with rings of volcanic ash and pumice. Later, erosion along the lakeshore cut into the crater margins, so the maar volcanoes are not very obvious now.

A little more than a million years ago, still in Pleistocene time, new lava flows spread over part of the present monument. These flows of gray basalt seem to have been related to similar flows in the Warner Mountains well to the east, but whether they actually came from those mountains is doubtful. They are exposed now in the northern part of the monument, particularly along Gillems Bluff. Some erupted from distinct fissures, such as the one called Big Crack near the northeast monument boundary. Though volcanic ash deposits occur with the gray basalt, the ash came from outside the national monument area.

Around 100,000 years ago, a shield volcano developed around a vent south of the present monument. Most of the volcanic landforms in the monument developed on the north slope of this shield volcano during three active eruptive phases:

• Initial cinder cone and lava flow activity created most of the cones you now see in the monument: Bearpaw, Caldwell, Crescent, Eagle Nest, Hardin, Hippo, Island, and Whitney Buttes. The eruptions were staggered in time through thousands of years, from 100,000 to 1,100 years ago, and covered most of the older shield volcano's lavas with mantles of cinders and new lava.

• Mammoth Crater activity occurred about 30,000 years ago, along a fissure in the southwest part of the monument. An outpouring of molten basalt pooled into a large lava lake and ultimately overflowed to inundate almost all of the national monument area. Most of the lava tubes are in these Mammoth Crater flows.

• Eruptions of the youngest cinder and spatter cones occurred more than 1,100 years ago: Cinder Butte, the spatter cones of Fleener and Ross Chimneys, and the spatter ramparts of Black Crater and the Castles. All of these features except the lava flows associated with Cinder Butte bear a thin white mantle of pumice known to have come from Glass Mountain, 15 kilometers (10 miles) south. Glass Mountain's eruption has been dated at 1,000 to 1,500 years ago. Only the lava flows associated with Cinder Butte are not dusted with Glass Mountain pumice. They are younger—their age has been dated at 1,110 years ago.

Where volcanism is so recent, one always wonders what the future holds in store. There is no reason to think that the underground fires have died. New lava may at any time break through to the surface, erupting as popcorn-sized pumice to build new cinder cones. Eruptions will probably be presaged by earthquake shocks and possibly by volcanic gases escaping from fissures in the ground. There is little likelihood that this area will see as sudden and as devastating an eruption as that of Mount St. Helens, as basalt lava normally erupts much more quietly.

OTHER READING

Macdonald, G. A. 1966. *"Geology of the Cascade Range and Modoc Plateau."* In *Geology of Northern California,* California Division of Mines and Geology Bulletin 190, pp. 65–96.

Mount Rainier National Park

Established: 1899
Size: 952 square kilometers (368 square miles)
Elevation: 573 to 4,392 meters (1,880 to 14,410 feet)
Address: Tahoma Woods, Star Route, Ashford, Washington 98304

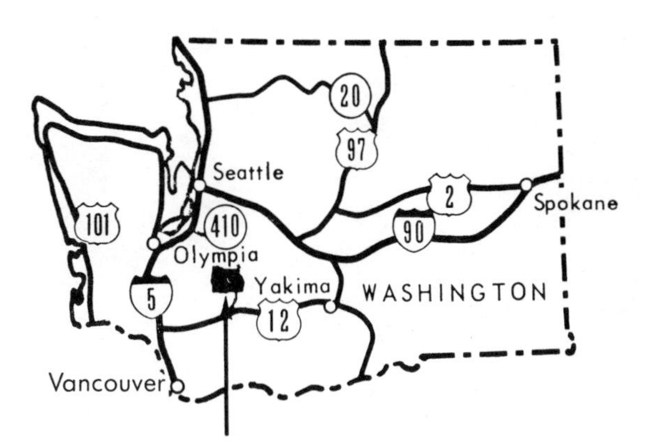

STAR FEATURES

• Mount Rainier itself, highest and most splendid of the volcanoes of the Cascade Range. The towering peak is an eroded stratovolcano once even larger than it is today.

• Glacier-carved ridges and peaks on which are exposed some of the many lava and ash flows that make up the mountain.

• Forty-one modern glaciers, visible from many vista sites. Born in summit ice fields, the largest of these reach down to elevations of about 1,200 meters (4,000 feet).

• Glacial features such as moraines, arêtes, matterhorns, U-shaped valleys, and hanging valley waterfalls that bespeak the great glaciers of Pleistocene time.

• Evidence of devastating mudflows and rock avalanches, results of important processes still working to reduce the mountain mass.

• Visitor centers, wayside exhibits, and a wealth of trails, some with self-guide leaflets. Naturalists lead excursions and present illustrated talks that help to explain many geologic features.

See cover and color pages for additional photographs.

SETTING THE STAGE

There is no sight more impressive than this majestic peak towering in ice-shrouded grandeur above mountains already 2,000 meters (6,000 feet) above the sea. The dazzling ice and snow of its summit spawn glaciers that reach long arms toward the lowlands, giving birth in their turn to foaming

Deeply crevassed glaciers cascade down icefalls and squeeze between narrow ridges of rock. In this photograph, taken in the summer of 1980, lower parts of the glacier are dusted with ash from eruptions of Mount St. Helens.

cascades milky with finely ground rock flour.

Mount Rainier is a stratovolcano, companion to Mount Hood, Fujiyama, and Vesuvius. Like those volcanoes, it has had in geologic terms a short but checkered career, often peacefully sleeping, at times waking suddenly to murmur and steam, now and then exploding in incredible violence. Its position in the path of moisture-laden winds from the Pacific guarantees the summit a perpetual cloak of snow and ice, a gleaming mantle that during parts of the Ice Ages extended to the lowlands and the sea. Here, surface features are regulated by a long-drawn battle between volcanic fire and glacial ice.

Today, 41 glaciers flow from high cirques and summit ice fields, following channels that radiate from the mountain crest. The largest of them are on the north and east flanks, where they are best protected from the sun's heat. In smoothly furrowed valleys the glaciers flow evenly—faster at the center, more slowly along the edges. Where the path is rough, they crack in thundering icefalls deeply scarred with crevasses. As they reach warmer elevations, the ice melts as fast as it is replaced from the frigid heights.

What makes a glacier? Snow, cold, and sloping terrain. On Mount Rainier's heights, much more snow falls in winter than can possibly melt during the summer. At first it lies lightly on summit and slope, turning gradually into tiny spheres of ice, the skier's "corn snow," the geologist's névé. New snowfalls compress the névé, squeezing out the air between grains of ice. As years pass, the ice thickens. When it is about 30 meters (100 feet) deep, the highly compressed, dense ice at the bottom becomes plastic, able to flow stiffly like thick taffy. Gradually it inches its way downslope, following pre-established stream courses as routes of least resistance.

Fed by yearly additions of new snow, the glacier grinds toward warmer elevations. There, eventually, it halts. But does it? Even near its lower end the ice still flows, moving with a patience that matches its own immensity, melting at its snout, replenished in its lofty birthplace. Like a conveyor belt, it moves and yet remains in the same place.

And like a conveyor belt it carries a burden, a cargo of rock and soil that it has plucked and gouged from its bed, or that falls to its surface from

The snout of Nisqually Glacier, darkened with ash from Mount St. Helens, constantly discards slabs of weakened ice.

bordering ridges. Using rocks and soil as the teeth of a giant rasp, the moving glacier chisels and scours, freezing to solid rock, plucking away large blocks, pulverizing them into fine white flour. Over the centuries it deepens and straightens and smooths its valley into the well known U-shaped trough that distinguishes glaciated mountains.

Mount Rainier's lava flows are separated by bands of volcanic ash and broken volcanic rock that appear as irregular, sloping layers in the walls of the glaciated valleys. The lava is dense and resistant; the once-airborne debris is soft and readily eroded. In places both have been weakened and discolored by hot groundwater and volcanic gases which change many of the volcanic minerals to brown and yellow, easily recognized, easily eroded clay. As individual eruptions poured lava from the summit crater, the flows, like the glaciers, sought easy valley paths down the mountainside. Each valley flow was long and slender, a spoke radiating from the mountain's hub. (A geologic map available at visitor centers illustrates this well.) Some spokes were 600 meters (2,000 feet) thick; some were many kilometers long. As incandescent lava and fiery blasts of hot pumice streamed across them, glaciers melted, and mudflow mixtures of ash, rocks, and meltwater sped down the valleys, streaming onto the lowland below and occasionally reaching even to Puget Sound, more than 80 kilometers (50 miles) away. Congealing lava and solidifying mudflows pushed streams aside, out of their own valleys and onto adjacent ridges. Continued downward cutting by these streams and, in times of glaciation, by reestablished glaciers produced new channelways separated by the resistant spokes of volcanic rock. Staggered eruptions later sent new flows down the new valleys, repeating the sequence. Evidence of these cycles can be seen today on lava-crested ridges and lava-coated benches far above the floors of present canyons.

Mount Rainier's gray volcanic rock is andesite, a type of igneous rock that contains a high proportion of plagioclase feldspar. Quite unlike the black basalt of the ocean floor, it formed from continental material. Extending from British Columbia to California, the Cascade volcanoes are near the edge of the continent, along a linear zone where dark, heavy basalt of the ocean floor slips downward beneath the advancing edge of North America. Along the line of junction, some of the granitic and sedimentary rocks of the continent are dragged down to a deep zone where remelting can occur.

Once melted, the lighter rock is more buoyant than its surroundings, and pushes surfaceward again much as oil drops rise through water, seeking pathways among fractures and faults or simply melting its way upward. Some of the magma, like the granodiorite that underlies Mount Rainier, cooled deep underground. But some broke through to the surface, perhaps as earthquakes opened new faults, to fuel the long line of the Cascade volcanoes. Molten rock poured out on the landscape, at times frothing into volcanic pumice and ash that erupted with explosive violence.

In volcanically active periods, the size of the mountain increased by additions of lava and ash. During periods of inactivity, erosion gained the upper hand. Here, we see the age-old conflict between mountain building and wearing down by erosion expressed as a conflict of fire and ice.

GEOLOGIC HISTORY

Cenozoic Era. Probably about 1.5 million years ago, in the middle of the Pleistocene Epoch, when continents of the northern hemisphere were locked in glacial ice, the first Mount Rainier lavas broke through to the surface. Prior to that time, however, earlier volcanic events left records in the rocks that now fringe the great mountain:

• Early in Tertiary time, 60 to 40 million years ago, a broad, slowly sinking shelf edged the western seaboard, a sea-covered platform that resembled the continental shelf off the south and east coasts of the United States today. Volcanoes on the submerged shelf emitted lava that piled up over a vast area to a total thickness of 4,500 meters (15,000 feet). Initially the volcanoes erupted under water, but as time went on they built up their own masses until at least some of them projected above the sea as islands. As sea water filtered into hot magma reservoirs, underwater steam explosions rocked these volcanoes, shattered the lavas, and caused enormous underwater mudflows that spread fine vol-

The rock of the Tatoosh Pluton is granodiorite, an intrusive igneous rock similar in appearance to granite.

canic material across the shelving bottom. The gray-green rock of these mudflows now makes up the Ohanapecosh Formation.

As volcanic activity abated, the region was lifted above the sea and then eroded almost to sea level again. At the time the region eroded, the climate in this part of North America was warm and humid; soils that developed on the eroding surface were the red soils of the tropics.

• Around 30 to 25 million years ago, after many millions of years of erosion, volcanoes outside the present park showered this region with volcanic ash and other debris. Most of the ash swept out as billowing plumes of rapidly moving, incandescent ash buoyed up by its own rapidly expanding gases. As it came to rest, the still-glowing particles fused into welded tuff recognized now as the Stevens Ridge Formation, light gray or white rock looking very much like sandstone but containing flattened, angular fragments of darker gray pumice.

• In another episode of volcanism, basalt and light gray andesite flows now known as the Fifes Peak Formation covered the northern part of the park. Dikes and sills of the same age occur in several widely scattered areas within the park.

Rocks of these three pre-Rainier volcanic episodes—the undersea breccia, the thick welded tuffs, and the basalt and andesite flows, dikes, and sills—were then lifted again and subjected to another period of erosion.

• About 13 million years ago, late in Miocene time, more molten rock pulsed upward from the depths. Though it compressed and broke the overlying rock, it never reached the surface. Instead, cooling slowly in a large magma chamber deep underground, it crystallized into the speckled granodiorite of the Tatoosh Pluton. (The word pluton means an igneous rock mass that cooled beneath the surface.) Easily distinguished by its black-and-white, salt-and-pepper texture, the rock of the Tatoosh Pluton probably underlies most of Mount Rainier.

One more round of uplift occurred after intrusion of the pluton: the wide-scale lifting and folding of the Cascade Range. As the land rose, erosion shaped a new surface, a rugged mountainous terrain slashed with valleys. Erosion bared the layers of volcanic rock and the previously hidden mass of intrusive granodiorite. The stage was now set for eruption of Mount Rainier.

In mid-Pleistocene time, a long chain of volcanic centers gave birth to many of the Cascade volcanoes. At first, lava welling up at these centers and reaching the surface flowed along erosional canyons, eventually filling them to their brims. Volcanic ash, rising in heated clouds or propelled sideways by violent explosions, showered down on the lava flows. Repeatedly, cycles of lava and ash, again and again carved by erosion, built up the volcanic cones. For a time volcanism dominated, and the cones increased in size. About 75,000 years ago, the volcano that is Mount Rainier reached its zenith, with a summit at about 5,000 meters (16,000 feet). Then volcanic action abated, perhaps only temporarily, perhaps permanently, and the mountain was whittled to its present size.

Even before the growth of the Rainier volcano, glaciers had formed in the Cascade highlands. Since the time of its maximum development, Rainier has at least three times been deeply sheathed in ice with glaciers extending from its flanks out onto surrounding lowlands. Most of the glacial features now evident on the mountain—the cirques, the steep-walled glacial valleys, the moraines—are products of the last 25,000 years. About 15,000 years ago, the glaciers began to shrink. Even in historic time Mount Rainier's glaciers have been shrinking, though since 1951 they have thickened and advanced just a little.

Since the shrinking of the glaciers, 22 recognizable layers of volcanic ash and cinder have blanketed the mountain slopes. Eleven of these came from Mount Rainier itself; they are thickest on the east side of the mountain and may not appear at all on the western, windward side. Ten are gray ash distributed on all flanks of the mountain, known to have come from Mount St. Helens. One, a thick yellow layer, came from the climax eruption of Mount Mazama in southern Oregon—the eruption that caused the steep-walled Crater Lake caldera. Several of the layers have been dated and can be used to delineate older and younger events. The Mount Mazama ash, for example, is known to be nearly 7,000 years old and is found over seven northwestern states and in three Canadian provinces. An admirable time marker, it separates layers older than 7,000 years from layers younger than 7,000 years.

Mount Rainier's most recent eruption occurred little more than a century ago, when it strewed its own uplands with pumice fragments that still lie like burnt popcorn on some surfaces. The mountain now is a dozing giant, giving off only enough heat to partly melt the summit ice and to warm occasional storm-bound climbers. We cannot tell when or whether it will awaken. Eruptions of Mount St. Helens remind us that the Cascade volcanoes are not extinct and provide a foretaste of the pandemonium to come when this larger volcano, with its much vaster ice fields, returns to life. Fortunately, Mount St. Helens provides a living volcanic laboratory where geologists can seek better

Avalanches and rockslides, powerful forces in erosion's arsenal, tear down the mountain that volcanism built.

methods for forecasting eruptions. Mount Rainier's glaciers, it must be remembered, store 8 cubic kilometers (2 cubic miles) of water in their frozen grip, water that hangs like the sword of Damocles above thickly populated lowlands. Mudflows from Mount Rainier could cause far greater tragedy than the tragic floods that ravaged the Toutle River near Mount St. Helens. There is clear evidence that an eruption once sent a cataclysmic surge of rock, mud, and meltwater from Mount Rainier all the way to Puget Sound. Obviously, only the ability to forecast eruptions in time for evacuation could save the thousands who live beneath Rainier's sword of ice.

BEHIND THE SCENES

Backbone Ridge. The crest of this ridge, where it meets the Stevens Canyon Road, follows almost exactly the contact between the Ohanapecosh Formation on the east and the Stevens Ridge Formation on the west. Both rock units are volcanic and older than Mount Rainier. The Ohanapecosh Formation is easily recognized by its gray-green color and the small, dark green rock fragments within it. The Stevens Ridge Formation is light gray, mostly welded tuff. Stone balustrades along the Stevens Canyon Road are made of rocks of these two formations.

Take a good look at the large rockslide on the west side of Backbone Ridge. In areas of rugged terrain, slides and rockfalls are some of erosion's most potent weapons.

Carbon Glacier. Carbon Glacier descends from an unusually large cirque whose high semicircular walls cut deeply into the volcanic rocks of Mount Rainier's north face. Leached and weakened by percolating hot water and volcanic gases, these rocks are easily undermined by moving ice. Frequent rockfalls keep the glacier covered with dark debris.

This glacier descends to a lower altitude than any other glacier in the lower 48 states. It can be reached easily by hiking about 3 kilometers (2 miles) up the Carbon River from Ipsut Creek Campground.

Cowlitz Box Canyon. A large Pleistocene glacier descending the Cowlitz Valley scraped away rock that lay above a resistant sill of intrusive rock here.

Glacial striae can still be seen on the scoured surface. After the glacier retreated, its meltwater sliced a narrow, twisting chasm through the sill, probably following a vertical joint or the shattered rock of a narrow fault zone.

Cowlitz Chimneys. Craggy remains of thick layers of volcanic breccia surround several volcanic plugs, once conduits for lava and breccia of the Ohanapecosh Formation. Rock forming the plugs is more resistant to erosion than the surrounding fine-textured material, and it projects as the sharp spires of South Cowlitz Chimney, Double Peak, Barrier Peak, and Bald Rock.

East Side Highway. Skirting the east slope of Mount Rainier, this road passes between the volcano and the crest of the Cascades. The route is mostly in gray-green volcanic rocks of the Ohanapecosh Formation. Many of the summits in this part of the park are strengthened by resistant dikes and sills, in most places too heavily timbered to be readily distinguished. They are exposed near Tipsoo Lake, where there are excellent views of Mount Rainier's distant summit and, closer at hand, the Cowlitz Chimneys.

Emmons Glacier. Best seen from viewpoints near Sunrise, this glacier can be closely approached by trail from the White River Campground. From its birthplace on Rainier's summit to its beveled snout, the glacier cuts deeply into the volcanic rocks that make up Mount Rainier. It is bordered by Steamboat Prow on the north and Gibraltar Rock and Little Tahoma Peak on the south—volcanic remnants that tell us Rainier was once even larger than it is now. What starts out as a single, broad ice field at the mountain's crest divides to form three glaciers: Winthrop Glacier north of Steamboat Prow, Emmons in the center, and Ingraham on the south. Emmons Glacier involves several streams of ice separated by knife-sharp ridges. Angular peaks (matterhorns) and narrow ridges (arêtes) are typical features of alpine glaciation, as are the moraines visible at the sides and lower ends of glaciated valleys.

Emmons Glacier's gray moraines are partly obscured now by a pink blanket of broken rock that resulted from rock avalanches on Little Tahoma Peak in 1963. Masses of rock tobogganned wildly down and across Emmons Glacier. Buoyed up by cushions of compressed air, they raced downvalley at speeds close to 150 kilometers (100 miles) per hour. They came to rest a short distance above White River Campground. The avalanches may well have been triggered by a steam explosion, as rangers 20 kilometers (12 miles) away heard a loud boom coming from this direction on that day. The quantities of fallen rock hide the ice surface of the glacier, and the scar they left on nearby Mount Ruth is clearly defined.

Kautz Creek Mudflow. A semifluid mass of mud and rock surged down Kautz Creek in October 1947

The beveled snout of Emmons Glacier is dark with debris from the 1963 avalanches. The rock may have been loosened from Little Tahoma Peak by steam explosions.

and partly buried and smothered the forest here. The mudflow began high on Mount Rainier's south slope, rushed down the mountainside carrying huge blocks of ice and rock, and for a time piled up in a narrow part of the canyon. Breaking through, it surged with redoubled strength into the lower valley.

At least six earlier mudflows have inundated this valley. They occur with sobering frequency on many of the Cascade volcanoes, where large quantities of ice and snow cling precariously to rock leached and weakened by volcanic gases and oversteepened by glaciers.

Lake George, Mount Wow, and Gobblers Knob. Lake George lies in a small glacial basin below the northern cliffs of Mount Wow. Thick erosional remnants of andesite and volcanic breccia belonging to the Ohanapecosh Formation make up both Mount Wow and Gobblers Knob. These rocks, the breccia in particular, contain unsorted but somewhat rounded volcanic pebbles and boulders cemented by a finer groundmass. They are well exposed along the upper part of the Gobblers Knob Trail.

The view from Gobblers Knob includes Sunset Amphitheatre and valleys of the Puyallup River and Tahoma Creek. About 2,800 years ago a rock avalanche of altered and weakened, yellowish brown volcanic rock broke away from the west slope of Mount Rainier and shot down the south Puyallup Valley as a mudflow. Its magnitude may be judged from the fact that it surged north over 250-meter-high (800-foot) Round Pass, near the parking area for the Gobblers Knob Trail, and into the valley of Tahoma Creek. A similar mudflow occurring only 600 years ago swept northwestward down the Puyallup River valley to the Puget Sound lowland.

A large glacier moving down Tahoma Creek in Pleistocene time carved away the east side of Mount Wow, leaving stunning vertical cliffs now adorned with slender waterfalls.

Longmire (elevation 841 meters, or 2,760 feet). The visitor center here houses displays explaining the geology of Mount Rainier. Rampart Ridge, sheltering the Nisqually River's valley, is capped with one of the long valley lava flows that became ridges when yet older ridges on either side eroded away. This flow is one of Rainier's longest. Its lower end is 15 kilometers (10 miles) from the volcanic hub of the mountain.

The Nisqually River winds across a flat-floored valley and among several highly mineralized hot springs once advertised for their supposed medicinal value. The hot and locally carbonated water shows us that volcanic fires still smoulder not far beneath.

Rock from Little Tahoma Peak (background, in clouds) shot across Emmons Glacier, leaped from its snout, and came to rest near the White River Campground.

Deep in a glacial cirque, Lake George is a halfway stop on the hike to Gobblers Knob.

Slender curving columns along the Puyallup River trail formed as a lava flow cooled and shrank. Columnar jointing such as this is common in volcanic rocks, but rarely is it as beautifully displayed.

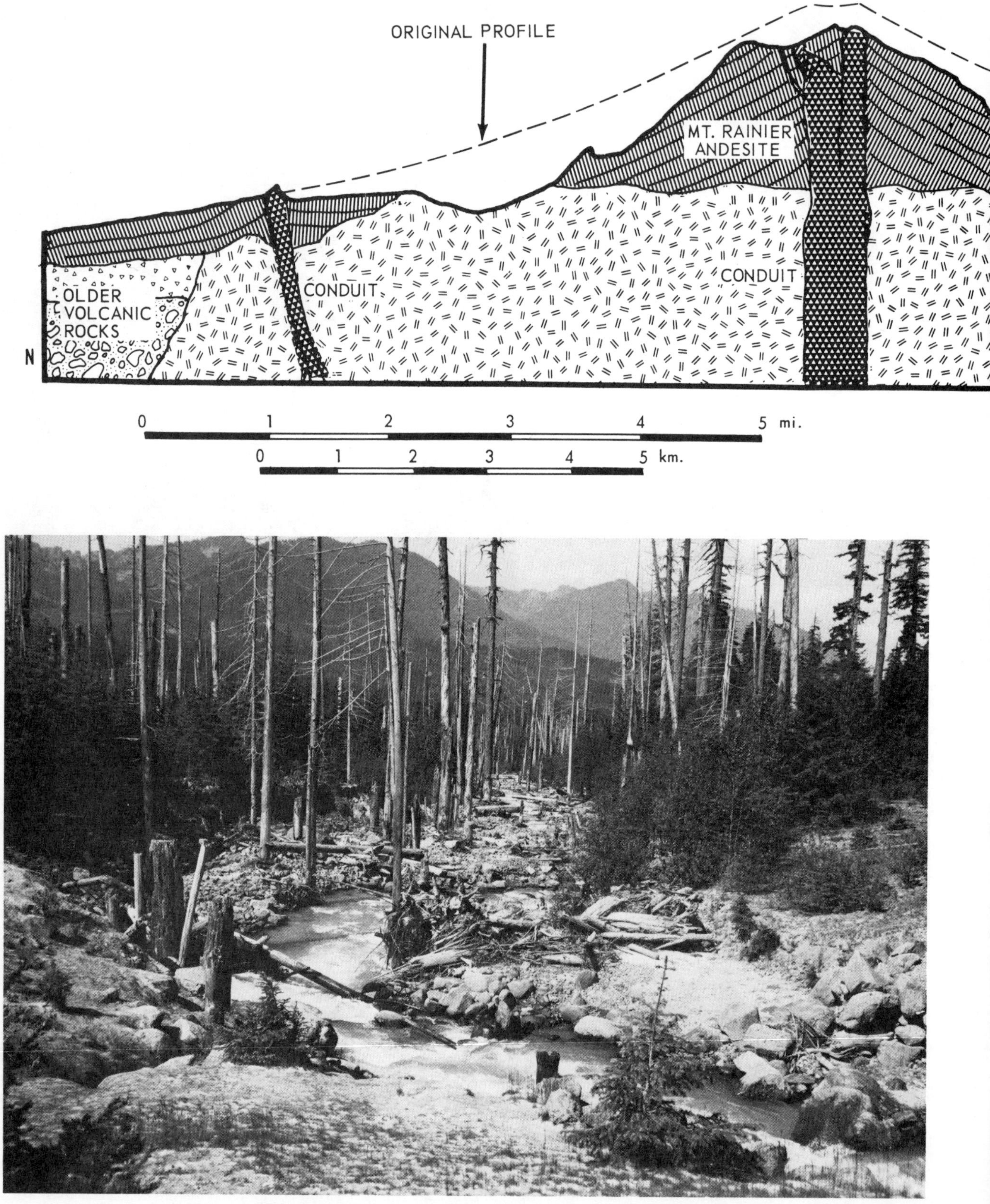

A recent mudflow down Tahoma Creek left a medley of rock, mud, and dying trees.

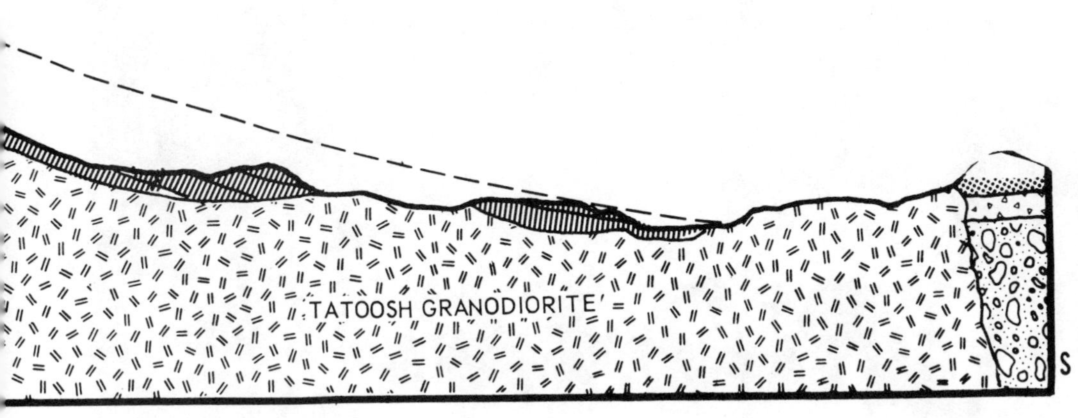

Youthful volcanic rocks of Mount Rainier lie on an eroded surface of older igneous rock. The profile of the mountain as it was about 75,000 years ago can be reconstructed from today's topography.

Mount Rainier Summit (elevation 4,392 meters, or 14,410 feet). The highest point on Mount Rainier is on the narrow rim between two relatively small, overlapping summit cones. Their rims usually blown clear of snow, these cones can be seen in profile from several eastern and southern vantage points. They are about 300 meters (1,000 feet) high, a little larger than Sunset Crater in Arizona (a national monument) and the same size as Cinder Cone near Lassen Peak (a national park). They developed 2,000 years ago in the calderalike depression of Rainier's summit.

Long before the May 1980 eruption of Mount St. Helens, geologists had postulated a remarkably similar eruption for Mount Rainier, in which an old summit, softened and altered by steam and other hot volcanic fumes, would be shattered by an explosion that would send a helter-skelter mass of clay, rock, and melted ice and snow rushing headlong down the mountainside and northward via the White River to Puget Sound.

The ice-covered summit still exudes some volcanic heat, enough to melt out ice caves under the summit ice cap and to reduce the ice cap enough to delay total erosion of the new summit cones. Occasionally steam eruptions are reported. No geologist would rule out new and possibly disastrous eruptions of this mighty volcano.

Despite the warmth at the summit, Mount Rainier is still deep in the frigid grip of its own private Ice Age. Seven major glaciers originate at its crest, others a short distance below.

Narada Falls. The salt-and-pepper rock at Narada Falls (best exposed in the unnatural "outcrops" of the parking area balustrade) are part of the Tatoosh Pluton, the large igneous intrusion that serves as Mount Rainier's foundation. The rock is granodiorite, a close kin to granite. Paradise River plunges over the contact between this resistant rock and two Mount Rainier lava flows. Downstream are several more falls bordered by the resistant granodiorite.

Nisqually Glacier. In the 1830s, this glacier extended below the site of the present highway bridge. During the last Pleistocene advance, it reached another 25 kilometers (15 miles) downstream. After many years of shrinking, the glacier is now advancing again; each year it pushes closer to the highway bridge. From the bridge you can see its high-walled valley and the jumbled rock debris left at its last retreat. Many of the rock fragments bear glacial scratches.

The glacier, deeply crevassed, flows between two straight-topped lateral moraines, features best seen from foot-trail vantage points near Paradise. The

D. L. Crandell photo

A rock glacier below the Palisades may still be ice-lubricated so that it flows much as a glacier does.

triangular faceted ridge above the western moraine shows a cross section of some of Rainier's volcanic flows, alternating lava and ash discolored over many centuries by steam and volcanic fumes. Like the flow that forms Paradise Ridge, those flows now stand up as a prominent ridge, though once they filled a valley. Someday perhaps another generation of lava flows will fill the present Nisqually Valley, melting the glacier and displacing the stream.

From Nisqually Vista or Panorama Point on Glacier Vista Trail, look down on the arcuate crevasses of the glacier. From those viewpoints both lateral moraines show up well, with the makings of a medial moraine in a row of rockpiles stretching lengthwise down the glacier. The canyon floor near the glacier is covered with debris that slid from the moraines. A tributary stream cascades across the western ridge. In regions of more resistant rock, this stream would plummet from a vertical wall of a distinctly U-shaped valley, but here the volcanic rocks have been so altered by steam and other gases that they collapse in rockslides.

Ohanapecosh River and Hot Springs. Most of the rivers of this park are white or gray with fine rock flour ground by moving glaciers. But the Ohanapecosh, heading in snowfields in the eastern part of the park, does not receive much glacial water, so it normally runs clear and emerald green. Near the campground, green outcrops of the Ohanapecosh Formation intensify its color.

The hot springs, used for a time as a spa, are now murky ponds surrounded by travertine deposits. Surface water sinking deeply into the ground, heated by contact with steam and hot rock, rises to the surface here, bringing with it calcium carbonate and other minerals in solution. As the water cools, these minerals precipitate in mounded cones of travertine. Slimy surfaces around the pools and channels are brightly colored with algae especially adapted to life in hot water.

The Palisades. The resistant rock of the Palisades is the welded tuff of an ash flow so hot that it fused as it came to rest. Vertical cooling joints similar to those in some lava flows give the Palisades a stockadelike appearance. The tuff has been dated as Miocene, older than the Mount Rainier volcanic rocks. Hummocky rock piles below the cliff wall originated little more than 10,000 years ago with the melting of a small, rock-covered glacier that had huddled below the cliffs. Ice flowage beneath and among the rocks may still be taking place. Rock glaciers like this one are numerous in near-glacial environments where there is a generous supply of rock debris.

Paradise (elevation 1,676 meters, or 5,500 feet). High on one of Rainier's old valley lava flows, this ridgetop area offers unparalleled views of "the mountain" and of the glaciers that course its southern slopes. Exhibits in the visitor center and along the trails explain geologic features.

From Nisqually Vista Overlook or the Skyline Trail at Panorama Point, one of the small summit cones is in view, as are the Nisqually Glacier's blunt snout and lateral moraines.

In clear weather the view south from Panorama Point or Alta Vista includes three more Cascade volcanoes: Mount Adams, cone-shaped Mount Hood, and Mount St. Helens. During some summers you can visit ice caves beneath the snout of Paradise Glacier, with a fascinating walk across a surface not long freed from glacial ice.

Stevens Canyon Highway. Most rocks along this highway predate Mount Rainier. East of Backbone Ridge are exposures of gray-green Ohanapecosh Formation; west of it are lighter gray or sand-colored flows of the Stevens Ridge Formation. At Cowlitz Box Canyon the road descends through these formations into older intrusive rocks that predate even the Tatoosh Pluton.

West of Sylvia Falls the road follows a deep, glaciated valley carved in salt-and-pepper granodiorite of the Tatoosh Pluton.

Sunrise Ridge (Yakima Park). The surface of this flat-topped ridge is sprinkled liberally with popcornlike pumice of an eruption that occurred about 2,000 years ago. In roadcuts along the edge of Sunrise Ridge, older layers of volcanic ash are exposed. The lowest is a sandy orange layer about 5 centimeters (2 inches) thick that has been dated at a little less than 7,000 years old. Its chemical composition tells us that it is the product of the cataclysmic final eruption of Mount Mazama, 400 kilometers

Three sets of fractures cut welded tuff along the Stevens Canyon Road.

The U-shaped valley of the Cowlitz River cuts into hard intrusive rock probably of Oligocene age.

(250 miles) to the south—the eruption that created the circular caldera of Crater Lake. Two other "foreign" ash layers, one of them 50 centimeters (20 inches) thick and about 3,500 years old, have been traced to Mount St. Helens, only 80 kilometers (50 miles) away. The 1980 St. Helens eruption, which deposited about 1.5 centimeters (less than an inch) on Mount Rainier's southeastern slopes, was puny by comparison.

A number of trails depart from the Sunrise

Columnar jointing in volcanic rock develops at right angles to the cooling surface. This example, along the road to Sunrise, is the very toe of a lava flow.

area, short ones to Emmons Glacier Vista and Sourdough Ridge, longer ones to Mount Fremont, Glacier Overlook, and Emmons Glacier itself. Any of these routes will provide excellent views of Rainier's east face, its magnificent glaciers and sharp arêtes, and profiled remnants of the once larger mountain.

Tatoosh Range. Though composed largely of the intrusion, or pluton, to which it gives its name, this range includes other rocks as well. On their steep north faces, both Stevens and Unicorn Peaks show volcanic layers of the Ohanapecosh and Stevens Ridge Formations, altered and in places brightly colored by later volcanic vapors and the heat of a small intrusion. Pinnacle Peak farther west is capped with isolated remnants of Stevens Ridge ash flows. The jagged range, deeply cut with small cirques, still harbors a few little bodies of ice, too small and stationary to qualify as glaciers.

White River and the Osceola Mudflow. In the upper White River Valley just below Emmons Glacier, the debris of 1963 rock avalanches is piled against the moraines of the once-larger glacier. The sea of debris is a reminder of another savage incident of the past—the Osceola mudflow, whose size far surpasses that of any other mudflow on the mountain. About 5,800 years ago, 2 cubic kilometers (half a cubic mile) of clay, silt, and boulders swept down the valley of the White River and across the Puget Sound lowland to the sea, a total distance of more than 80 kilometers (50 miles). If a similar mudflow were to occur today, the present communities of Enumclaw, Buckley, Kent, Auburn, Sumner, and Puyallup would lie in its path.

Where did all this mud come from? How was it able to flow so far? Geologists have found scattered

About 5,800 years ago the Osceola mudflow poured a thick, churning mush of rocks, mud, and timber into the valley of the White River. A similar mudflow swept down the Puyallup River valley just 600 years ago. Tephra *is a geologic term that includes volcanic ash and coarser airborne debris. It is mapped only where it is more than 2.5 centimeters (1 inch) thick.*

Adapted from Crandell, 1973

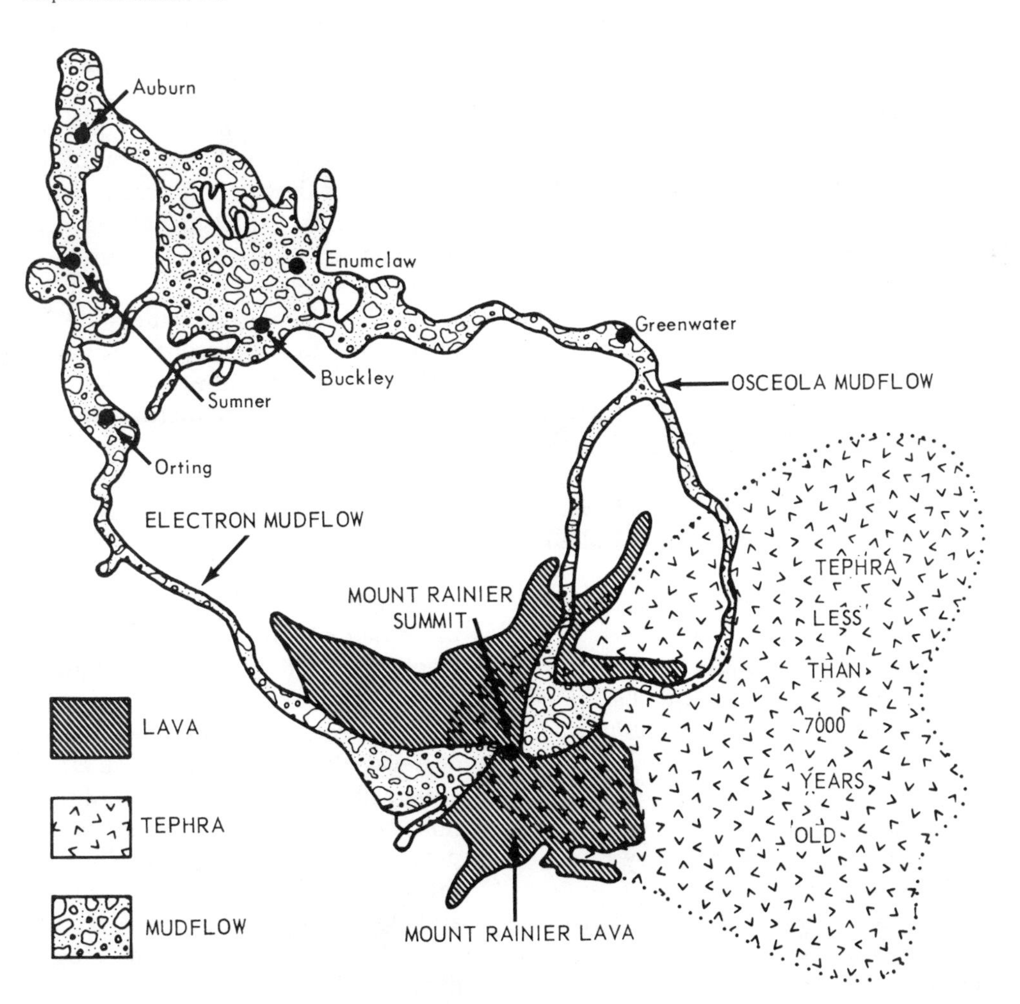

patches of the mudflow at the top of Steamboat Prow and on the ridges that enclose Glacier Basin, so it must have originated somewhere above these features on the ice-covered upper slopes of Mount Rainier. There is strong likelihood that the mudflow was the former summit of the volcano, demolished by an explosive eruption. Destruction of the summit left a deep amphitheatre which has since been partly filled by the two summit cones.

Probably the volcanic rock of the old summit was saturated and highly altered by steam and volcanic vapors, and probably the exploded material, mixed with new volcanic ash and searing gases, rapidly melted the summit glaciers. The altered rock would account for the high proportion of fine clay that lent mobility to the mudflow, as well as for the weakening of the summit that caused it to yield readily to explosive forces. And melted glaciers would explain the huge volume of water that carried summit materials to such a great distance.

Wonderland Trail. Though this trail encircles the mountain (an eight- to ten-day hike), it touches the highway at many points and so is equally accessible for shorter hikes. Zigzagging over ridges topped with resistant lava flows, descending into ice-carved valleys, it leads past many of Rainier's outstanding volcanic and glacial points of interest. Hikers should watch for faults and joints, pumice-covered uplands, exposures of lava and ash flows (some with columnar jointing), dikes and sills, volcanic plugs, lakes and waterfalls caused by glaciation or erosion of hard and soft rocks, moraines, and glacial striae.

OTHER READING

Crandell, D. R. 1969. *Superficial Geology of Mount Rainier National Park.* U.S. Geological Survey Bulletin 1288.

Crandell, D. R. 1969. *Superficial Geology of Mount Rainier.* U.S. Geological Survey Bulletin 1292.

Crandell, D. R. 1973. *Potential hazards from Future Eruptions of Mount Rainier, Washington.* U.S. Geological Survey Map I-836 (with text).

Crandell, D. R., and Mullineaux, D. R. 1967. *Volcanic Hazards at Mount Rainier, Washington.* U.S. Geological Survey Bulletin I-238.

Fiske, R. S.; Hopson, C. A.; and Waters, A. C. 1963. *Geology of Mount Rainier National Park, Washington.* U.S. Geological Survey Professional Paper 444.

Fiske, R. S. 1964. *Geologic Map and Section of Mount Rainier National Park, Washington.* U.S. Geological Survey Map I-432.

Mount St. Helens National Volcanic Monument

Established: 1982
Size: 445 square kilometers (172 square miles)
Elevation: about 250 to 2549 meters (about 820 to 8364 feet)
Address: Gifford Pinchot National Forest, 500 W. 12th Street, Vancouver, Washington 98660

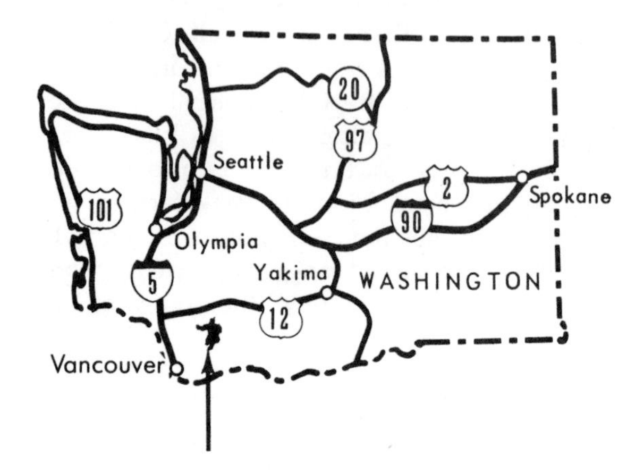

STAR FEATURES

• An active volcano in the Cascade chain, dynamic, up-to-date evidence of the Plate Tectonic Theory in action.

• Access routes providing good views of the volcano, of Spirit Lake, and of mudflow damage and widespread destruction occasioned by its May 18, 1980 eruption. Roadside displays describe many geologic features; park interpreters give talks at major viewpoints.

• A visitor center near Interstate 5, five miles east of Castle Rock, presents explanatory exhibits, movies, and slide shows; smaller information centers dot other nearby highways.

See color section for additional photographs.

SETTING THE STAGE

On the morning of May 18, 1980, in a sudden paroxysm of fury, Mount St. Helens exploded. Geology in action! As an earthquake loosened an immense rockslide, an explosion shot rock debris and new volcanic material northward in a scenario of devastation. Though a large eruption had been anticipated, on the basis of continuing small eruptions, persistent earthquake activity, and the appearance of an ominous, rapidly growing bulge on the volcano's north flank—the size of the eruption and the searing, suffocating, rock-laden lateral blast that left 57 people dead or missing, were not predicted.

First recognition of Mount St. Helens' growing potential for violence came on March 23, nearly two months before the great eruption. In the preceding few days, seismometers at the University of Washington had recorded sudden strong earthquake activity beneath the volcanic mountain—activity confirmed by additional seismometers quickly put in place on the mountain itself.

Over the next few weeks these instruments were to record hundreds of additional shallow quakes, many of them greater than magnitude 4, many in clusters typical of pre-eruption patterns recorded in Japan. At this stage, the public was asked to stay away from Mount St. Helens and the Spirit Lake area to the north. Areas above treeline were closed, largely because of the possibility of earthquake-triggered avalanches.

Small eruptions of steam and ash began on March 27, with small new craters forming on the mountain's snowy crest. Though most of the ash fell on the mountain itself, some dusted nearby communities. Earthquakes and small eruptions continued to make news. Residents and loggers living or working within 24 kilometers (15 miles) of the mountain were urged to leave. Scientists stepped up their observations, expanded their sampling programs. They found that the ash was pulverized older rock—the stuff of which the mountain was built—rather than new volcanic material, and that most of the gases were steam derived from groundwater heated by contact with hot rock.

Steam and ash eruptions increased in April. Earthquakes became stronger, though less frequent. Some quakes generated harmonic tremor,

Photographs do not do justice to the devastation wrought by the May 18, 1980 eruption of Mount St. Helens. As in this picture, taken from Windy Ridge, the horseshoe-shaped crater is often wreathed in swirling, windblown volcanic ash.

USGS photo, courtesy of David A. Johnston Cascade Volcano Observatory

The volcano's May 1980 eruption darkened skies more than 500 kilometers (300 miles) downwind. Hot ash melted the mountain's crown of glaciers, creating many mudflows on its slopes. Wispy vertical streaks below the ash clouds are raining ash.

which produces an undulating pattern on seismograph records. Tremor of this kind suggested motion of fluid magma somewhere beneath the volcano. Again, monitoring was intensified, both on the ground and in the air. Volcano hazard warnings were issued, and roadblocks were set up to control access. Nevertheless, many sightseers, hikers, and climbers slipped into the controlled area; some even reached the summit crater.

A new and frightening element came into the picture toward the end of March. Cracked ground and new crevasses in glaciers showed that the volcano's north flank was swelling, bulging outward because of pressures within the mountain. As the swelling grew with the passing days and weeks, officials worried about the possibility of massive avalanches. Conceivably, an earthquake-triggered landslide might shoot down the mountainside to Spirit Lake, causing a surge in lake water that would suddenly and with little warning flood the Toutle River valley. Daily information releases after April 23 notified area residents and visitors of this growing peril.

Earlier, U.S. Geological Survey geologists had studied this and other Cascade volcanoes to ascertain the perils posed by their snow-capped summits and the incandescent depths below them. With a long history of explosive eruptions, Mount St. Helens was described as an especially dangerous volcano, the most likely of all the Cascade array to erupt. When? "Possibly before the end of the century." Studies showed that previous eruptions, spaced out over 4500 years, had involved many bursts of volcanic ash, pumice, and rock fragments, with massive mudflows generated as hot ash melted snow and glacial ice.

Like its near neighbors Mount Adams and Mount Rainier, Mount St. Helens is a stratovolcano, composed of irregular, alternating layers of lava and of the once-airborne rock debris and volcanic ash that characterize explosive eruptions. Its pre-1980 summit was shrouded in snow and ice; a dozen glaciers radiated from its crown. The mountain itself stands a little west of the main parade of Cascade volcanoes, a row which stretches from British Columbia to northern California, marking the line where the East Pacific Plate plunges steeply beneath the continent. There, molten magma, much or most of it derived from light-weight and light-colored continental material, pushes up through the continental crust, fueling the line of volcanoes.

Throughout early May, earthquakes continued. They were monitored by a growing array of instruments, some automatic, some manned, some on the mountain's slopes, others at surrounding vantage points. Daily flights photographed small new

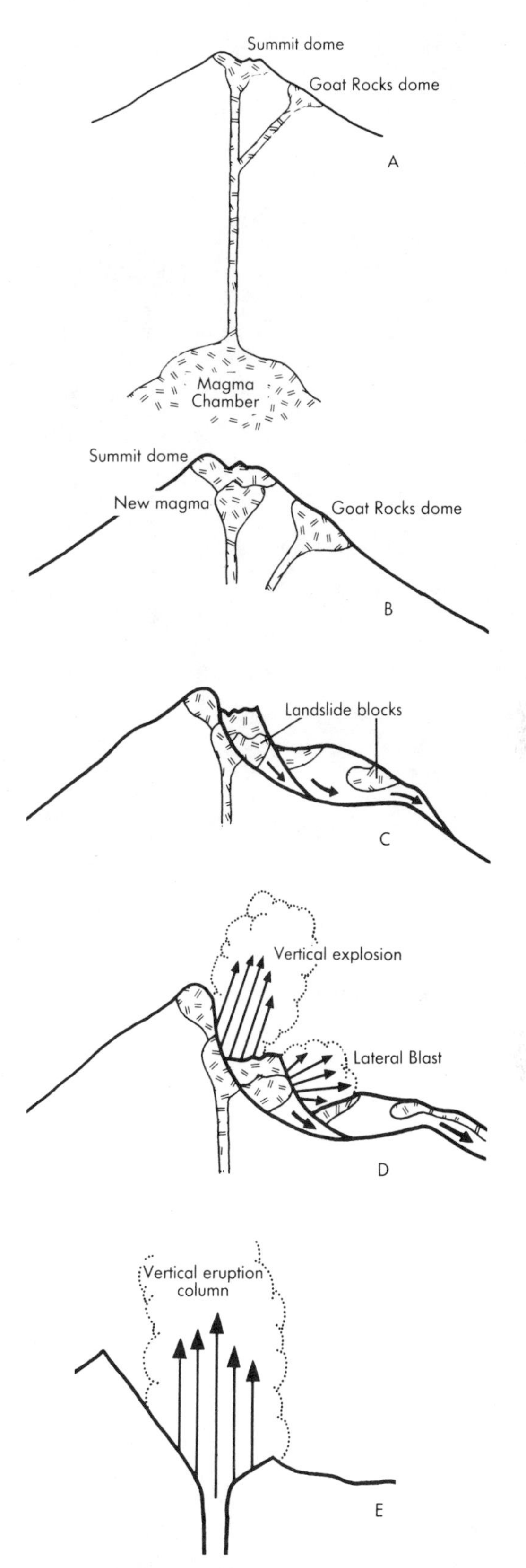

Adapted from USGS Earthquake Information Bulletin v. 16, no. 2

Triggered by an earthquake, the May 18 landslide in turn triggered the violent lateral explosion of Mount St. Helens.

Seven miles from the explosion site, shattered tree trunks stand among the debris of their neighbors. Fine gray ash covers the ground.

summit craters and steam vents and sampled gases emitted from them. Gas samples were also obtained directly from these vents, the last on May 17. Sulfur compounds among the volcano's gases showed that molten volcanic rock was not far below the surface.

Meantime, the ominous bulge area on the north slope continued to grow at a measurable rate of 1.5 meters (5 feet) a day. The bulge now appeared to be highly cracked and wet with melted snow. Warnings were issued, but many ignored them. On May 17 some residents were allowed to enter the "Red Zone" under escort, to recover possessions from homes and cabins.

Then, at 8:32 in the morning of May 18, an earthquake struck, magnitude 5.1. The unstable bulge began to move, plummeting northward down the slope, taking with it the very summit of the mountain. Breaking up as it descended, the massive avalanche of rock and ice shot northeast into Spirit Lake, rode up the ridges north of the mountain, or veered westward down the valley of the North Toutle River.

Removal of so much material from the mountain slope and summit released the pressures that had for so long held in the volcano's gargantuan power. Hot on the heels of the rock-ice avalanche, a volcanic blast raked the land with pulverized rock. The searing blast, its temperatures reaching an estimated 360°C (680°F), traveled at speeds as great as 325 meters per second (725 miles per hour). Its rapid advance snapped off trees, flinging them over ridges like so many giant jackstraws; farther from the explosion site it laid them in neat windrows across successive ridges. Forest, animals, and some 57 people—fishermen, sightseers, volunteer volcano-watchers, a few loggers, and a Survey geologist stationed on a high ridge west of Spirit Lake—died instantly. The blast obliterated everything within a fan-shaped wedge 37 kilometers (23 miles) broad, extending 29 kilometers (18 miles) north of the original mountain summit.

The lateral blast, lasting only a few minutes (but registered as a high-pressure atmospheric wave in Maryland and New York some three and one-half hours later), was prelude to other events. For nine hours the volcano was in continuous eruption. Black clouds boiled upward in a column 24 kilometers (15 miles) high, hiding the volcano. As their dark pall spread eastward with the wind, gritty,

The Toutle River mudflow carried destruction many miles from the scene of the explosion. Note the high-water mark on trees at left.

powdery gray ash fell thickly over Washington, Idaho, and western Montana, and in diminishing amounts farther east across the nation. Hot, ground-hugging pyroclastic flows swept northward from the hidden crater, depositing a thick apron of pumice across the blast-denuded landscape. Lightning flashed from the roiling, highly charged ash clouds, lighting forest fires. Water from melted ice and snow rushed down the mountainside, gathering momentum, feeding mudflows that swept north and west down the two forks of the Toutle River and southeast down the Muddy River. The flow from the Toutle—thick mud, like wet cement, so dense it floated boulders—ultimately reached the Cowlitz River, which carried its debris to the Columbia. Bridges along the Toutle were ripped out by the mudflow; they would be sorely missed by rescue workers in the days to come.

Though the cataclysmic eruption of May 18 ended abruptly at 5:30 in the afternoon, for three more days the volcano spat forth intermittent ash clouds. Then, as earthquake activity diminished and the clouds cleared away, it was discovered that the top of the mountain and its northern slope were gone, and that in their place a yawning, horseshoe-shaped amphitheatre rose starkly above a vast, gray sea of desolation. Spirit Lake—once so beautiful—was a steaming morass floating innumerable dead trees, their bark and limbs stripped away completely. Floods diminished, but forest fires burned on. Rescue missions brought out survivors—many of them with vivid, harrowing tales to tell. Towns and cities dug their way out of the ashes.

Since May of 1980 there have been other eruptions of steam and ash, and of light gray dacite lava too thick to flow, accumulating in lava domes within the crater. Domes built in June and August of 1980 were later blown apart in explosive eruptions. A larger lava dome, which made its debut in October, 1980, was built in thirteen increments and survives, 250 meters (800 feet) high, as of this writing (September, 1984). Except for small amounts of steam and ash, the volcano has been quiet for the last year. Earthquake activity has lessened. But Mount St. Helens is still considered dangerous. The potential is still there. Spirit Lake, dammed by the debris flow from the collapsing bulge, is still a major threat. Present lake level is 80 meters (260 feet) higher than the original lake level. To prevent it from rising, overtopping, and washing away its new dam, lake water is being pumped through a newly constructed tunnel into the North Fork of the Toutle River.

Monitoring continues, and every effort is being made to give warning should the volcano threaten another major eruption. The Mount St. Helens disaster has brought with it new knowledge, knowledge that can be applied to predictions of future eruptions of this and other stratovolcanoes. Since the May 1980 eruption, U.S. Geological Survey scientists working with up-to-the-minute data have successfully predicted many smaller eruptions several hours or several weeks in advance. With warning, risks and damage in surrounding areas may be minimized.

The most important data for short-term predictions come from seismic monitoring and measurements of ground deformation. Thirteen seismometers, one of them right in the crater, record earth tremors originating in, under, and around the volcano, even down to the slight tremors caused by rockfalls, avalanches, and gas bursts from the lava dome in the crater. (Larger quakes are also recorded farther afield, by seismometer networks in place well before the 1980 eruptions.) Observations have shown that shallow earthquakes under the dome build up gradually in the days before each new dome-building eruption, and then increase dramatically just a few hours before the eruption begins. Measured tilting and swelling of the crater floor or of the great dome within it usually give the first indication of impending activity. Other data—the appearance of wrinkles, radial cracks, or small thrust faults on the crater floor, and changes in gas emissions—are also important portents of coming eruptions.

GEOLOGIC HISTORY

Cenozoic Era. In this area, Precambrian, Paleozoic, and Mesozoic rocks are hidden beneath thick layers of volcanic rock. Our geologic history begins, therefore, at the beginning of the Cenozoic Era, about 65 million years ago. At that time, you'll remember, the North American Plate began to drift westward, riding over the East Pacific Plate. In mid-Tertiary time, along the zone where the two plates overlapped, fissures developed and molten magma pushed upward from the Earth's mantle.

On many hillsides trees fell like wheat before a scythe. By 1984, when this picture was taken, lupine and pussytoes (lower left) bloomed in sheltered spots.

USGS photo, courtesy of David A. Johnston Cascade Volcano Observatory

The surface of Spirit Lake is now many meters above its former level. Here it reflects the new Mount St. Helens and the dark dome in its horseshoe-shaped crater.

The earliest magma was basalt, derived from the mantle. It erupted gently, non-explosively, and flooded the land repeatedly, slowly building up the Columbia Plateau of Washington and Oregon.

Not until about 15 million years ago did the character of the erupting magma change. Then, over many millions of years, volcanic outpourings became lighter in color and more silicic in composition—evidence that they were no longer derived purely from the mantle, but were made of continental material drawn down and melted along the subduction zone at the edge of the continent. Since these new silicic magmas were thicker and stickier than the earlier basalt magmas, and did not flow as freely, they piled up into a stately chain of cone-shaped mountains—the Cascade volcanoes.

Although the history of Mount St. Helens begins about 37,000 years ago, the more recent volcanic story of the mountain dates back about 4000 years ago. By that time the terrain around the volcanic center was not much different than it is today: Stream courses had been established, though they would be modified as the mountain grew. The Columbia River was where it is now, entrenched in the layered lava flows of the Columbia Plateau. Probably the great northwestern forests were every bit as green as they are now. There had been no volcanic activity at the present site of Mount St. Helens for at least 4000 years.

Then, after the 4000-year pause, eruptions began—explosive bursts of gas, steam, and rock fragments and quieter eruptions of thick, pasty, dome-forming lava. Dark clouds of volcanic ash surged skyward; other ash clouds, buoyed up by the hot gases within them, shot outward from the volcanic center. Lava domes built up and then were blown to bits by later eruptions, their fragments contributing to the growth of the volcano. Through the years, the mountain grew higher. When it had become tall enough to wear a crown of snow and ice, eruptions of hot ash sent mudflows coursing down its flanks into the valley of Smith Creek, Ape Creek, and the two forks of the Toutle River.

Intermittent eruptions continued for almost 2000 years, depositing more broken volcanic rock, more ash, more mudflows, and occasionally more lava flows on the flanks of the growing volcano. Between 400 and 1400 A.D. the eruptions lessened—

On the southeast side of the volcano, melted ice and snow thick with volcanic ash and broken tree trunks became the Muddy River mudflow.

U.S. Geological Survey scientists measure gases emanating from slopes within the crater. Increasingly sulfurous gases may reflect magma movement, which in turn indicates an approaching eruption.

USGS photo, courtesy of David A. Johnston Cascade Volcano Observatory

Streams and rills are cutting new dendritic (tree-like, branching) drainage patterns in the volcanic ash.

a lull that would not last. Since 1400 A.D., eruptions have been relatively frequent, every 100 years or so. They have created, over and over again, the same types of volcanic ash and fragmented rock, the same associated mudflows, as were later emitted by the 1980 eruption. Several lava domes built up, some of them soon exploded or eroded away. Many lava flows, as well as pyroclastic flows of hot gases, ash, and rock fragments, swept downslope, adding to the mountain's volume. Right up to a mid-19th-Century cluster of eruptions, when a lava dome—Goat Rocks—appeared on the volcano's northern slope.

In 1978 the U.S. Geological Survey, which had been studying the Cascade volcanoes to ascertain the hazards they posed, pointed out that Mount St. Helens was an especially dangerous mountain, with a history of frequent explosive eruptions. Maps were drawn up showing areas that had been covered by lava, pyroclastic flows, and mudflows during the last 4000 years—areas where the volcano could strike again. The mountain was watched. The warning signs came. However the portents—earthquakes, steam and ash plumes, the ominous swelling bulge—did not forewarn of the nature of the explosion or of the magnitude of the lateral blast. Yet without them, and without already established evacuation procedures, hundreds, perhaps thousands more people would have died.

Large in its human context, the May 1980 eruption seems smaller in its geologic context. Many greater eruptions are known, a few of them in historic time. In 1912 Mount Katmai in Alaska exploded with a blast about ten times as great as the initial blast of the St. Helens eruption—fortunately over an uninhabited area. Krakatau (Krakatoa) in Indonesia blew itself apart in 1883, with an explosion heard 3200 kilometers (2000 miles) away, showering neighboring islands with glowing boulders and generating tidal waves that killed more than 30,000 islanders. It disgorged seventy times as much volcanic material as Mount St. Helens. Mount Mazama, which 6900 years ago stood where Crater Lake does today, spread similar volumes of ash over its surroundings and indeed over much of western United States, and then collapsed into its own magma chamber (see Crater Lake National Park). Thera, in the eastern Mediterranean, in one terrifying instant destroyed the entire thriving Minoan civilization.

Except during periods of impending volcanic activity, Mount St. Helens National Volcanic Monument is as safe to visit as any mountain area. Danger areas, in a ring around the mountain itself, are well marked, and the volcano is under constant surveillance. New roads traverse the stricken forest and approach good viewpoints outside the marked danger areas. Bridges have been rebuilt across the great mudflows. Earthquakes and other warning signs are studied with care; if they suggest that an eruption is likely, the area will of course be cleared immediately. The many small information centers keep up with the status of approach roads and with Geological Survey predictions of volcanic activity.

OTHER READING

Foxworthy, Bruce L., and Hill, Mary. 1982. *Volcanic Eruptions of 1980 at Mount St. Helens: The First 100 Days.* U.S. Geological Survey Professional Paper 1249.

Harris, Stephen. 1980. *Fire and Ice: the Cascade Volcanoes.* Pacific Search Press, Seattle.

Palmer, L., and KOIN-TV Newsroom 6. 1980. *Mt. St. Helens—The Volcano Explodes!* Northwest Illustrated.

Ream, L. R., and Jackson, Bob. 1980. *St. Helens! The 1980 Eruptions.* Bob Jackson Books, Renton, Washington.

Shane, Scott. 1985. *Discovering Mt. St. Helens, A Guide to the National Volcanic Monument.* University of Washington Press, Seattle.

Tilling, Robert I. 1985. *Eruptions of Mount St. Helens: Past, Present, and Future.* U.S. Geological Survey Publication.

North Cascades National Park

Established: 1968
Size: 2,042 square kilometers (788 square miles)
Elevation: 213 to 2,782 meters (700 to 9,127 feet)
Address: Sedro Wooley, Washington 98284

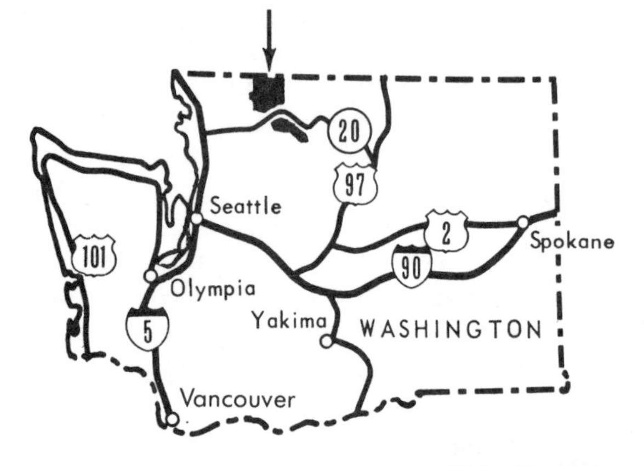

STAR FEATURES

• A rugged alpine wilderness shaped by past and present glaciers. Access to park lands is almost entirely by trail. Parts of the range can be explored in adjoining recreation areas and wilderness.

• Glacial features such as cirques, moraines, matterhorns, U-shaped valleys, and rock surfaces marked with glacial striae.

• Present-day glaciers, particularly on the western side of the mountains, where the high peaks intercept Pacific moisture.

• Evidence of mining days when gold fever brought prospectors into the rugged and then unknown hinterland.

• An abundance of trails, guided and self-guided walks and hikes, slide programs, talks by naturalists, and wayside displays, particularly in adjacent recreation areas.

Intricately contorted gneiss, once a layered sequence of sandstone and siltstone, reveals the intensity of pressures and temperatures that created these mountains.

SETTING THE STAGE

Elsewhere in the Cascades, old metamorphic and intrusive rocks that make up the foundation of the range are hidden beneath thick piles of volcanic rock. Here in northern Washington, however, the foundations are clearly exposed. Bent and broken rocks jut skyward as dark peaks, jagged ridges, and isolated uplands creased by deep glacial troughs.

The crags and peaks of the North Cascades are composed of a medley of rocks that defy simple description. They have puzzled geologists for decades. But thorough and careful mapping—not an easy task where dense forests clothe and conceal the surface—has revealed irregular northwest-to-southeast bands of sedimentary, metamorphic, and intrusive rocks, steeply and repeatedly contorted and faulted. In places, slices of rock are pushed over one another along thrust faults as if the mountains pushed out eastward on the east side, westward on the west side of the range.

Rocks in the heart of the mountains are banded gneiss and glistening mica schist that once were sand and mud, marble that once was limestone, colorful greenstone that once flowed out on the sea floor as lava, and hard white quartzite that may have been volcanic ash. All these metamorphic rocks have been subjected to great heat and pressure. They are now intensively folded and faulted. You'll see them at roadside, lakeside, and trailside, often standing on edge. They are particularly well exposed along the North Cascade Highway between the north and south units of the park. Where their many fractures, or joints, provide access for water, they may be shattered by frost.

Masses of granite and other intrusive rocks occur on the two sides of the range, separating the metamorphic core from less disturbed sedimentary rocks to the east and west, outside the park.

In the high country, small glaciers and snowfields cling to the rocky crags, and ice fields cap a few summits. But these bits of snow and ice pale to insignificance when compared with the great valley glaciers that shaped these mountains and excavated the long U-shaped valleys that now cleave the range.

When winter snow exceeds summer melting and snow patches increase year by year in size and

M. H. Staatz photo, courtesy of USGS

Closely spaced parallel joints allow water to penetrate massive intrusive rock, laying the way for freeze-and-thaw weathering.

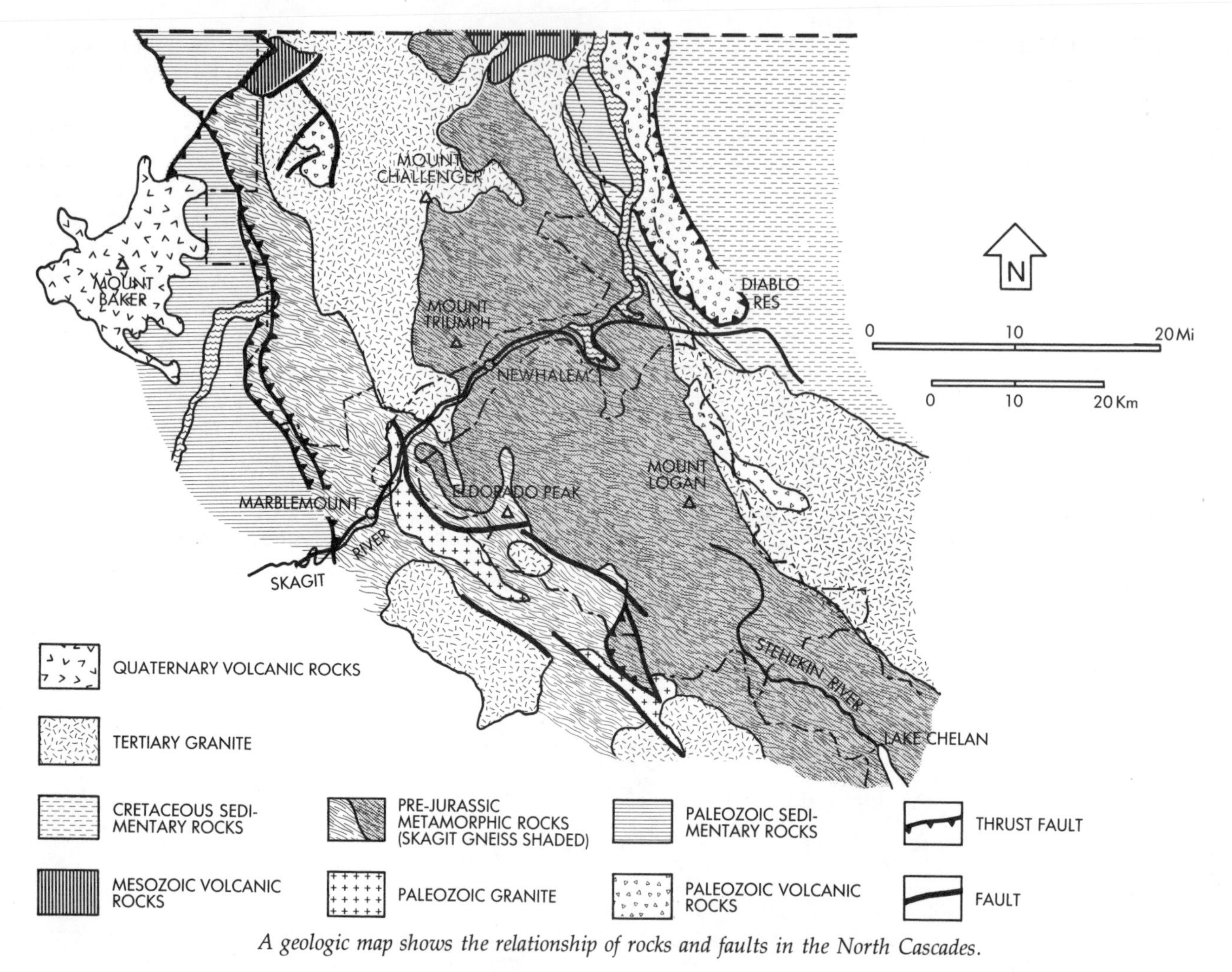

A geologic map shows the relationship of rocks and faults in the North Cascades.

thickness, the stage is set for development of glaciers. If snow and ice accumulate to depths of about 30 meters (100 feet), the lowest, well compacted layers of ice begin to flow. This is the moment at which an ice field, stationary until then, becomes a glacier. Moving downvalley from its birthplace among the peaks, the ice uses soil and rock like a giant rasp to grind and polish and shape its pathway. Replenished each year by new snow at the top, extending farther and farther down the valley, the glacier eventually reaches a point where addition of new snow at the top is balanced by melting of its lower end. Like a conveyor belt, the ice is still in motion, though the glacier as a whole seems stationary. Sometimes, after remaining stable for a time, the glacier begins to advance again, usually in belated response to periods of heavier snowfall in the high country. At other times, if the climate becomes warmer or precipitation decreases, it retreats. Through both advance and retreat, the ice continues its downhill motion.

What Earth forces lifted these mountains to our view? Probably the westward drift of the continent and the collision of the continent with one of several small, separate minicontinents that lay to the west. Just as a sheet of dough will bend and break if it is pushed across a rough surface, the Earth's crust here has crumpled and cracked. Parts of it were drawn downward and subjected to high temperatures and high pressures. Molten rock in places rose up through the cracks, filled the crumples, and, while still deep underground, hardened into granite. Eventually erosion removed sedimentary rocks that were not strengthened by metamorphism, bared the granite and the metamorphic mountain roots, and carved the North Cascades as we know them today.

GEOLOGIC HISTORY

Paleozoic Era. The early history of the Cascades is one of repeated upwarp and downwarp: sinking and rising land, advancing and retreating sea. Where this area lay then we have no way of knowing. It may have been much like the New Guinea-

M. H. Staatz photo, courtesy of USGS

Small glaciers and snowfields cling precariously to frost-sharpened peaks of the Picket Range, in the north unit of the park.

M. H. Staatz photo, courtesy of USGS

Some of the granitic intrusive rocks of the Cascades include blocky fragments of dark diorite from an earlier intrusive episode.

Sedimentary rocks broken, fused, and broken again are clues to the varied history of the North Cascades.

Cobbles of a mountain stream reflect the history of the region upstream. These, in Cascade Creek, are composed of metamorphic rocks, some of them crisscrossed by several generations of quartz veins. The veins originated as fluids that distilled from subterranean magma and penetrated fine rock cracks.

Celebes-Borneo part of the southwest Pacific today—island microcontinents awash in shallow seas. In Paleozoic time, deposits of marine limestone, sandstone, and shale covered and concealed thick Precambrian sedimentary layers, counterparts of the metasedimentary rocks now exposed in Glacier National Park.

About 350 million years ago, in Ordovician time, these rocks were broken to some extent, altered by heat and pressure, folded and forced upward into mountains. Then erosion leveled the mountains, and sedimentation around and across their site continued off and on past the end of the Paleozoic Era and into Mesozoic time.

Mesozoic Era. About 150 or 140 million years ago, during the Jurassic Period, a second generation of mountains appeared here. These again were folded and faulted ranges, their structure made more complex by intrusions of granite magma. And after they formed, erosion again carved them down. By the beginning of the Cretaceous Period, 136 million years ago, little trace of them remained. Once more the mountain roots were buried, this time by the sand and mud of a Cretaceous sea.

During the Mesozoic Era, though, deep-seated changes were taking place all over the world. Convective patterns in the semifluid mantle seem to have shifted about. Stresses caused by new patterns of convection rifted the great supercontinent, and segments of it began to drift apart. One of the segments became North America.

Cenozoic Era. As sea-floor spreading broke Pangaea apart, the Atlantic Basin widened and the Atlantic Ocean grew from a mere crack in the crust—a fault zone—to a narrow seaway similar to today's Red Sea, and then to the broad and still expanding ocean that we know now. The North American Plate embarked on its westward journey, riding out over the heavier and apparently stable East Pacific Plate. The forces involved—the friction, the resistance of the East Pacific Plate as it was forced downward under the continent—caused basic changes in the shape of the west half of the continent. The drifting continent encountered on its voyage the small Juan de Fuca Plate, actually one of three such plates added to the western edge of North America. The oceanic part of this plate, thin but heavy, sank before the onslaught, subducted along an oceanic trench. Granitic and sedimentary areas became part of the continent. In the melee, heat and pressure brought about metamorphism. Masses of granitic magma, perhaps formed from the melting of continental rocks dragged under in the collision, surged up into the weakened crust, and the North Cascades were born.

Folding and faulting in the Cascades seems to have reached a climax about 35 to 30 million years ago, possibly the time that the Juan de Fuca microcontinent was added to North America.

In late Pliocene time, as recently as 10 or even 5 million years ago, upwarping of a broad area (still powered by the continent's westward drift) raised the roots of the folded, faulted mountains and lifted them high above their surroundings. They were of course immediately attacked by erosion, and eventually the several-times-altered mountain rocks, with their granite intrusions and long-concealed metasedimentary rocks, were uncovered. Then, as erosion continued, volcanic eruptions created Mount Baker, Glacier Peak, and off to the south the stately parade of other Cascade volcanoes. As Mount St. Helens has so vividly demonstrated, this process, which began about a million years ago, is still going on.

Ice Age glaciation played a major part in designing today's spectacular alpine scenery here. During three (or possibly four) Pleistocene glacial advances, long rivers of ice born among the peaks of the North Cascades flowed from mountaintop to lowland plain. A particularly long, narrow glacier shaped the sinuous trough of Stehekin Valley, following and straightening a former river valley. Others ground out the valleys of Diablo and Ross Lakes (both artificial reservoirs now) and of Baker and Shannon Lakes west of the park. Still others crept down Little and Big Beaver Creeks, Goodell and Chilliwack, Panther and Thunder Creeks, and others. In all these valleys one can recognize the U-shaped profiles and hanging-valley tributaries that characterize glacial sculpture. At the valley heads are cirques from which the glaciers came, like scoop marks in ice cream mountains. The cirques are surrounded by tapering frost-sharpened matterhorns and keen-edged arêtes. On valley walls bare rock is marked with glacial striae, and on valley floors there are smoothed and rounded rocks geologists call, after their counterparts in the Alps, roches moutonées (sheep rocks).

As the valley glaciers developed, an expanding ice sheet from Canada pushed south into the Puget Sound Lowland. Slender tongues of this ice sheet pushed *up* the lower Skagit Valley and other west-flowing valleys of the North Cascades, damming their streams. By the end of Pleistocene time, the mountains were sharp and angular, honed by frost, rising tall above the valleys, which were not finally abandoned by the glaciers until about 8,000 years ago.

Born during a glacial advance about 4,000 years ago (long after man's arrival in America), today's small glaciers cling to rocky walls or nestle in

Near Cascades Pass a small glacier leads a perilous existence. Ice from its lip tumbles to cone-shaped piles far below. Slender waterfalls plummet into the valley of Cascade Creek.

cirques carved by their larger predecessors. Some are expanding, after a hundred years of melting back. Perhaps they presage a new Ice Age.

BEHIND THE SCENES

Cascade River Road and Cascade Pass. After traveling through dense rain forest on the west flank of the North Cascades and then turning up the North Fork of the Cascade River, this road enters the park only in its last few kilometers. However, it provides close-up contact with many of the rock types common in the park, notably the silvery mica schist that, because of the way it flakes and breaks, is responsible for much of the jaggedness of the summit ridges and peaks. Here, you can reach by road a mountain fastness otherwise accessible only by hiking.

Beyond the end of the road, the Cascade Pass Trail leads deeper into the park. Both road and trail saw their beginnings as Indian pathways. They knew the trudge of prospectors and miners of the last century and early parts of this one, and they were for a time the main access route into the Skagit River gold country. Gold was found, too, near the Cascade River's fork, where the road swings sharply left into the canyon of the North Fork. Marble was once quarried near Marble Creek.

At the end of the road, stop and look around. Walk up the trail toward Cascade Pass for a closer acquaintance with the precipitous cliffs and near-vertical snowfields. Many ice fields cling to the cliffs across the valley on the shoulders of Mount Chavall and Mount Buckindy; several rank as glaciers. On Eldorado Peak to the north, Inspiration and Boston Glaciers spawn tongues of ice that break off at their margins and tumble into steep ravines. Elsewhere, the rocky slopes are scarred with avalanche chutes, for valley walls oversteepened by glaciers are precarious holds for winter snow.

Lake Chelan. There is a Grand Canyon under the waters of Lake Chelan. The lake occupies an ice-carved gorge 2,600 meters (8,500 feet deep), its floor well below today's sea level. Were it not for moraine dams at its lower end, the valley might be a sea-filled fjord.

The long, finger-shaped lake, with its sheer walls, lies outside the park in Lake Chelan National Recreation Area, administered with the park. Along the lake shore are the rocks of which the North Cascades are made: granite intrusions of Cretaceous and early Tertiary age near its southern end; faulted, folded, and metamorphosed Paleozoic sedimentary rocks along its western shore; granite and Precambrian metamorphic rocks around its north end.

The glacier that carved the lake basin flowed down the Stehekin Valley from deep in the heart of the range, originating in the big cirque of Horseshoe Basin near Cascade Pass.

North Cascade Highway and Ross Lake National Recreation Area. Though not in the national park proper, this highway and recreation area are very much a part of the North Cascade scene. They provide access to many of the trails leading into the north and south units of the park. Highway cuts reveal the metamorphic and intrusive rocks of which the mountains are formed: the strongly banded gneiss crossed by coarse pegmatite veins and garlanded with branching streamers of white quartz; or west of Ruby Creek and east of Newhalem, the light-colored quartzite and salt-and-pepper granite. Rugged glacier-carved, frost-honed peaks and pinnacles tower above the highway; they, too, reveal the rock of which the mountains are made, but they are not easy to get to. Steep ravines—many of them avalanche chutes—widen joints inherent in this rock and occasionally harbor narrow, half-hidden waterfalls. Other falls tumble as thin veils from hanging valleys that border the larger, deeper valley of the Skagit River.

M. H. Staatz photo, courtesy of USGS

Banded gneiss displays variations from black to almost white, depending on the amounts of black minerals (biotite and hornblende) relative to white (quartz). The rock probably originated as alternating layers of sand and mud.

Skagit Tours will introduce you to the hydroelectric power plants powered by the waters of Ross and Diablo Lakes, both artificial reservoirs.

Stehekin Valley. Remote and untrammeled, this valley must rank as one of the most beautiful in North America. The deep trough was shaped in Ice Age time by a glacier coming from the region of Cascade Pass. Feathery waterfalls plunge from hanging tributaries once occupied by smaller glaciers unable to match the downcutting ability of the main Stehekin glacier.

At many places in this valley are rocks rounded by glacial action and marked with glacial striae. Moving ice rounded the upstream ends of rock bosses and quarried their downstream ends. When there is any doubt as to the direction of movement of bygone glaciers, such rocks are useful indicators.

For almost its entire length, the valley is walled with Precambrian igneous and metamorphic rocks. Some of the quartz veins cutting these rocks contain gold, especially above the wide-floored confluence with Flat Creek.

OTHER READING

Misch, O. 1977. "Bedrock Geology of the North Cascades." In Brown, E. H., and Ellis, R. C. (editors), *Geological Excursions in the Pacific Northwest,* Western Washington University, Bellingham, Washington.

Post, A., and others. 1979. *Inventory of Glaciers in the North Cascades, Washington.* U.S. Geological Survey Professional Paper 705-A.

Behind Mount Carrie, the summit of Mount Olympus lies swathed in glacial ice. Glacier-sharpened peaks and ridges mark both mountains.

Olympic National Park

Established: 1909 as a national monument; 1938 as a national park
Size: 3,677 square kilometers (1,420 square miles)
Elevation: Sea level to 2,428 meters (7,965 feet)
Address: 600 East Park Avenue, Port Angeles, Washington 98362

STAR FEATURES

• A snow-capped and long-enigmatic mountain range little touched by the hand of man. The Olympics are young mountains made of sedimentary and volcanic rocks deposited on the Pacific Ocean's floor.

• Tumbling cascades and quiet rivers, pathways from mountain to sea, that reveal in their canyons the geologic complexity of the range.

• Some 60 small glaciers, and evidence of past glaciation on a far greater scale.

• A wild, beautiful coastline with its own particular story.

• 190 kilometers (120 miles) of road and almost 1,000 kilometers (600 miles) of trails, the latter leading into otherwise inaccessible parts of the Olympic Mountains.

• Interpretive facilites and activities that include displays at three visitor centers, naturalist-led walks and hikes, self-guide trails, trail maps, numerous roadside and trailside displays, and illustrated talks.

See color section for additional photographs.

SETTING THE STAGE

As the northwesternmost outpost of conterminous United States, the Olympic Peninsula juts northward and seaward between Seattle and the Pacific. Here rise the clustered peaks of the Olympic Mountains, snow capped, glacier carved, often shrouded in clouds. Few roads lead into them; those that do cling to mountain ridges or wind up deep valleys that radiate like spokes of a wheel from the heart of the range.

The east side of the Olympics is fairly dry, with less than 50 centimeters (20 inches) of annual moisture. Outcropping rocks can be clearly seen. On the west, rocks wear mantles of moss, and soggy rain forests cloak geologic features.

The mountains are composed of sedimentary and volcanic rock, geologically young but severely bent and broken. Deciphering their short, intriguing history has not been easy. Early mapping, much of it in dense vegetation, revealed a horseshoe of volcanic rock embracing a chaotic central mass of crushed and twisted sedimentary rock layers lacking in fossils that might reveal their age.

The volcanic rocks are basalt, and in many places they are "pillow" basalt, with bulbous shapes known to form when molten lava erupts, cools, and hardens under water. Elsewhere there is volcanic breccia, a cemented-together mass of broken volcanic rock. These rocks resemble more than anything the submarine basalt now known to form along mid-ocean ridges.

Submarine lava bulges into rounded or barrel-shaped pillows. Here, the basalt pillows are tilted nearly 50 degrees from their original position.

Early workers decided that the mountains were the eroded remains of a large dome-shaped anticline, with the oldest rocks—as in all anticlines—at its center. They declared that the basalts were Eocene, 54 to 38 million years old, and that the sedimentary rocks, older than the volcanic horseshoe, were Cretaceous, 136 to 65 million years old. They were right about the basalt—it is Eocene—but wrong about the sedimentary rocks and the structure of the range. Careful study of ridges and cliffs and valleys and painstaking examination of the central mountains finally turned up some microscopic but datable fossils in the sedimentary rock. The sandstone and shale, ocean-floor sediments, are younger, rather than older, than the volcanic rocks.

Thin shale and sandstone layers, tilted 90 degrees, are visible on Hurricane Ridge along the trail to Hurricane Hill. Individual layers are 1 to 10 centimeters (0.5 to 4 inches) thick.

There were more enigmas. The sedimentary rocks are in many places tilted vertically, standing on end. And the layers go on, and on, and on. Are they truly so thick? Or are they tightly folded? Or repeated by faulting? The answers to these questions came slowly. To recognize tight folding, one must be able to distinguish the tops and bottoms of the layers. This can be done by studying the shape of ripple marks and the direction of graded bedding, in which coarse grains settle out first, so that grain size in any individual layer becomes finer upward. Gradually, as more and more field work was done, the picture became clearer. The layers

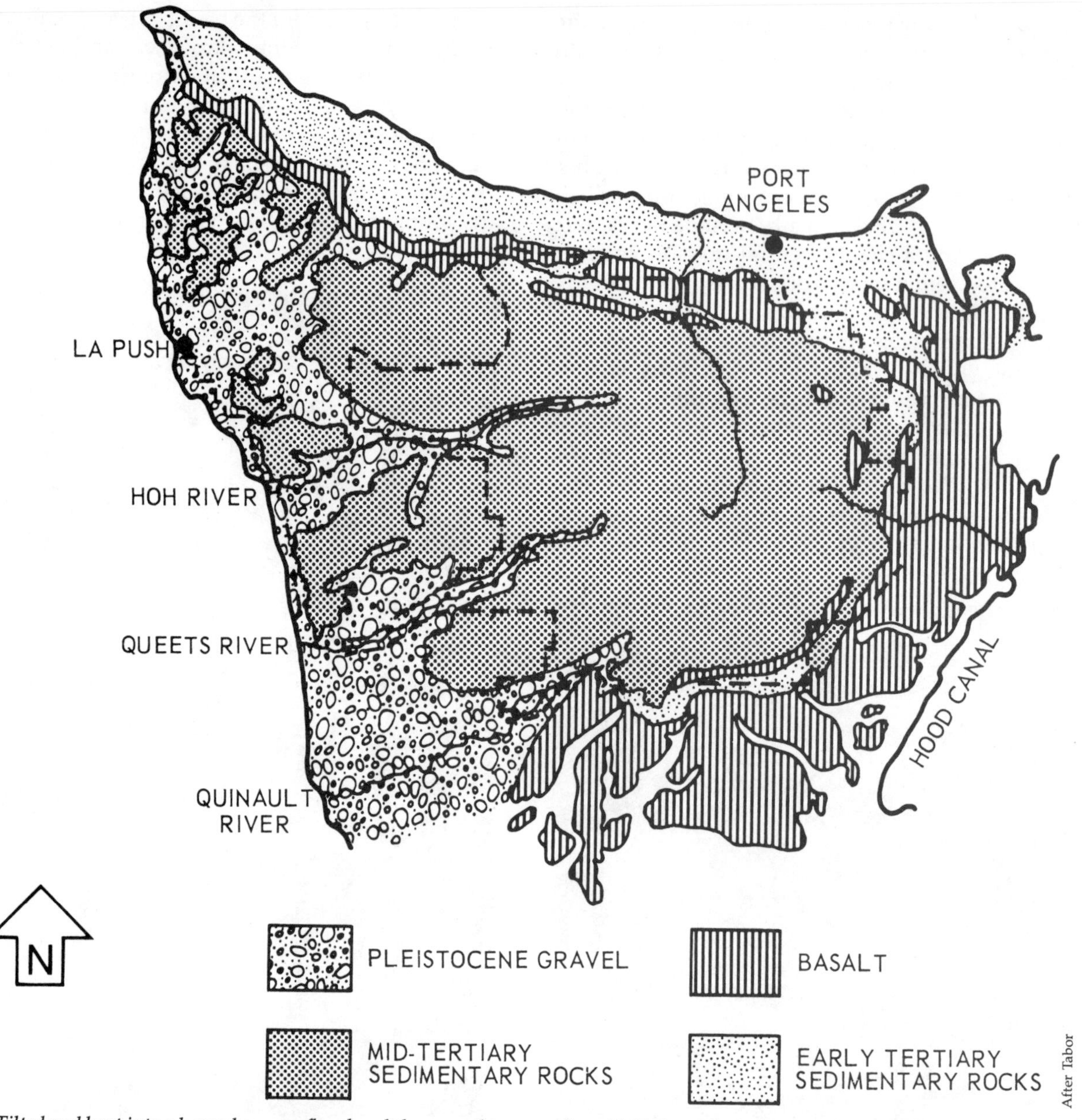

Tilted and bent into a horseshoe, sea-floor basalt layers embrace a region of folded and faulted sedimentary rock scraped from the floor of the sea. Stream-deposited glacial gravel mantles the Pacific coastal area and its major river valleys.

were folded—even accordion pleated—and they also were repeated by faulting. Certain characteristic layers could be identified more than once, and a few faults could be mapped. Most of the faults parallel the bedding, as if the sedimentary rocks were cut and piled upon each other like pancakes, tilted, and shoved hard against the volcanic horseshoe. Geologists now recognize more or less concentric bands of sedimentary rock, progressively younger westward, lying inside the arcuate bands of volcanic rock. Despite the folding, the youngest sedimentary rocks are those farthest west; the oldest are next to the volcanic horseshoe.

Until recently, the tremendous forces that must have been at work here were little understood. What pressure, coming in from the sea, could break these layers, stack them, bend them, tip them on end? Only with the coming of the Theory of Plate Tectonics did the baffling Olympic picture begin to make sense.

According to this theory, you'll remember, when the Atlantic Ocean Basin began to open, the North American Plate embarked on a westward migration that put it on a collision course with the East Pacific Plate. In the head-on meeting, the North American Plate came out, literally, on top. Composed of relatively lightweight granitic and sedimentary rocks, it slid over the East Pacific Plate's heavy basalts, which were forced down into the Earth's hot mantle. As the North American Plate journeyed over the

East Pacific Plate, its leading edge scooped up thick slices of pillow basalt and submarine breccia, as well as a portion of young sea-floor sediments, and piled them up like so many pancakes along the edge of the continent, in effect adding to its width.

In the process (which, incidentally, is still going on at the breakneck speed of a few centimeters per year), some of the sliced sedimentary and volcanic material of the old sea floor got shoved into an angle between what are now Vancouver Island and the Cascade Range—both older parts of the westward-drifting continent. Jammed into this aperture, the slices were tilted vertically and bent into an arc, just as our stack of pancakes might tilt and bend if we tried to push it into a gap between the coffeepot and the sugar bowl. The basalt layers, on the bottom, became the volcanic horseshoe of the Olympic Mountains. The slices of sedimentary rock, tilted vertically or even overturned, filled in the center of the horseshoe.

Many of the rivers flowing radially from the center of the Olympic Mountains are clouded with fine rock flour ground by the 60 glaciers that nestle in high cirques. A few of these glaciers are several kilometers long, among them Blue Glacier on Mount Olympus. The Olympics are not as high as the Sierra Nevada or the Rocky Mountains, nor are they as far north as the heavily glaciated ranges of Canada and Alaska. They owe their glaciers to their location on the edge of the continent, in the path of moisture-laden winds from the Pacific. Snowfall may exceed 60 meters (200 feet) a year here. Since more snow falls in winter than melts in summer, it accumulates until it is deep enough to compact into ice. And where ice reaches depths of 30 meters (100 feet) or more, the mere weight of it causes it to flow downhill following already established stream canyons, deepening them, widening them, straightening them into the U-shaped valleys that characterize glaciated regions.

Today's glaciers are small fry compared with the large valley glaciers that swept from mountaintop to sea during Pleistocene time. And those in turn were dwarfed by a vast river of ice that crept south out of western Canada, shaped Puget Sound, bumped into the Olympics, and split in two to course west to the Pacific and south around the east side of the Olympic Peninsula.

The effects of glaciation are not as marked in the Olympics as they are in Glacier, Grand Teton,

Sea caves are the first step in the process that creates sea stacks.

Stream-deposited gravel undermined by a wave-cut notch funnels down to a tidal bench. This process, too, may isolate sea stacks.

Rocky Mountain, or Yosemite National Parks, and for a good geologic reason. There the rocks are strong intrusive and metamorphic rocks and well hardened, billion-year-old sedimentary rocks. Here the rocks are young and not really well consolidated: sandstone and shale less than 54 million years old, in thin, brittle, easily separated layers, or volcanic rocks in rounded pillows that can be broken from their comrades without difficulty. Glaciers rapidly straightened, widened, and deepened narrow, twisting stream valleys. After the glaciers melted, unsupported valley walls collapsed in landslides and rockfalls or yielded to the accumulated soil creep of the centuries.

The coastal strip of Olympic National Park seems unrelated, except by its wildness, to the mountains. There, splendid sand and pebble beaches alternate with bold yellow headlands. Precipitous cliffs, caves, arches, and rocky islets rise above a wave-cut terrace, a shallow bench of solid rock that extends far out to sea.

Storm waves that break against the coast beat the cliffs with untold fury. With every wave, tons of water are hurled against the rock with a force great enough to compress the air in rock cavities, to explode the rock itself, to blast blocks of it bodily from the cliff. Surging back and forth across the rock bench, grinding it with tools of stone, the water carries its loot back out to sea or deposits it as pebbles and sand in sheltered spots along the shore.

Waves working thus create special landforms. At the base of cliffs they carve wave-cut notches, undermining the cliffs. They reach into every hollow to ream out sea caves. On headlands, two opposing caves meet to form an arch. Seas surging through the arch enlarge it until it collapses, isolating the tip of the former headland as a sea stack—a cliff-walled, often forest-crowned islet once part of the mainland.

Every wave washes sand and pebbles shoreward, then carries them seaward again, back and forth endlessly but with a year-to-year equilibrium. Even on calm days they winnow sand from the headlands and move it along the shore, building and changing the beaches. Here on the Olympic coast, waves meet the shore from the northwest, causing a net southward movement of both water and beach material. At river mouths or entrances to bays, the sand and pebbles pile up as slender spits or bars, partly or completely sealing off quiet lagoons.

High beach ridges change with every big storm. Over the long run, the beaches are backing away from the sea as headlands collapse and beach ridges push landward.

Examine the sand with a magnifying lens if you have one. Notice the equidimensional grains, the result of endless rolling by wave and wind. The sand has some unusual components here: dark, heavy, iron-bearing minerals like magnetite and ilmenite sorted out along swash marks on the beach, fine gold concentrated in this black sand by the surging sea, sorted grains of pink garnet that add a blush to Ruby Beach, miniature agates, grains and sometimes pebbles of black jade.

On the shore, and more visibly on some of the offshore islands, there are several terrace levels, old wave-cut platforms now above the sea. They tell us that this is a rising coast and that someday the present wave-cut bench, too, may be raised above the sea.

GEOLOGIC HISTORY

Geologically speaking, the Olympics are young mountains made from young rocks. Except for boulders and cobbles brought from Canada by a now-vanished glacier, there are almost no rocks of Precambrian, Paleozoic, or Mesozoic age in the national park. So let's start our discussion at the end of Mesozoic time, 65 million years ago, keeping in mind that North America had just broken away from Europe and struck out on its own.

Swash marks on the sand show concentrations of heavy iron-bearing minerals.

Cenozoic Era. The North American Plate continued to drift westward in Cenozoic time, overriding the Pacific Plate at the rate of about 5 centimeters (2 inches) per year. In the process it scraped enough sedimentary and volcanic rock off the Pacific Plate to add to itself a broad new band of land area.

The process was not as simple as the telling of it. When sediments and volcanic rocks of the sea floor were scraped up by the advancing continent, they broke and crumpled and folded and piled up sideways, as we have seen, like a stack of pancakes. Massive sandstone layers that had been deposited by underwater currents simply shattered, to be glued together again as groundwater rich in silica filled the breaks with white quartz. Siltstone and shale layers slipped, slid, and crumpled. Thick, sea-floor pillow basalt broke; thinner basalt layers bent with the sediments. Most of these processes took place far below the surface, where heat and pressure squashed sand grains and pebbles, altered shale into slate and slate into phyllite, and recrystallized volcanic layers to greenstone. Most of the fossils that surely must have been enclosed in the sedimentary rocks disappeared or were altered beyond recognition.

With the exception of a tiny bit of pre-Tertiary gabbro at Point of Arches, a product of a mid-ocean ridge now nearly swallowed up by the continent, no bedrock in Olympic National Park is older than the early Tertiary basalt of the volcanic horseshoe. The sedimentary rocks inside the horseshoe range from late Eocene to late Miocene, about 40 to 12 million years old, with just a few strata tentatively dated as Pliocene, 12 to 3 million years old. All the distortion—the bending, folding, and faulting—took place after those strata were laid down.

Sometime late in Tertiary time the Olympic area thrust upward, rising domelike from the sea. Streams ran in all directions off the high point of the dome, establishing a predominantly radial drainage pattern that still exists. There is evidence in the Olympics that there were at least two stages of uplift, with erosion reducing the first-stage mountains to rolling hills. Streams rejuvenated and strengthened by the second uplift cut into these hills, leaving a few smooth relict uplands to contrast with the rough mountain terrain.

At the beginning of Pleistocene time, about 3 million years ago, erosion sped up as the world's climate became cooler and wetter. Alpine glaciers, some of which may already have existed on the mountain heights, crept down stream-cut valleys. With their burdens of rock debris for tools, they scraped away soil and talus and landslide, and gouged into solid rock to create steep-walled cirques and to deepen and reshape the valleys.

Originally deposited on the sea floor as fine mud, this highly fractured slate now forms stony fields supporting little vegetation.

Evidence in the Olympics tells of at least four advances of these alpine glaciers and suggests that the central part of the mountains may more than once have been completely concealed beneath its own ice cap.

At the same time, a huge ice lobe from the continental ice cap that covered western Canada expanded southward through the Puget Sound lowland. Six times this lobe advanced as far as the barrier of the Olympic Mountains. There it split, with one part turning westward through the Strait of Juan de Fuca to the sea, the other diverted east and south around the Olympic Mountains and through the valley of the present Hood Canal. Telltale granite boulders from the ranges of western Canada were left as glacial erratics high on the flanks of the Olympics, showing that the ice lobe was at times nearly 1100 meters (3,500 feet) thick.

About 13,000 years ago, the climate became quite warm, even warmer than at present. The glaciers receded and finally disappeared completely. The 60 small glaciers present now in the national park are quite recent, having formed as the climate cooled again within the last 3,000 or 4,000 years.

As ice caps and glaciers disappeared the burden they had placed upon the land was lifted. Regions

Above and left: From its circling whirlpool the Elwha River thunders through the narrows of Goblins Gate, confined by hard, steeply dipping quartzite and slate that form the walls of Rica Canyon.

bowed down by the weight of glacial ice began to rebound. The west coast was rising anyway, so even though sea level rose as water locked within the world's great glaciers finally melted, broad river flats and deltas of river-deposited glacial debris rose faster. Along the shoreline, benches leveled by stormy Pleistocene seas rose with them, testimony to the flexibility of the Earth's ever-mobile crust.

BEHIND THE SCENES

Discussions in this section zero in on geology near the roads or within easy hiking distance of them. Most of the beaches discussed are also those accessible by road; others show variants of the same features.

Dosewallips River. The dry side of the Olympic Mountains displays many excellent outcrops of pillow basalt of the volcanic horseshoe. Thousands of meters of these rocks, near what must once have been the center of a substantial submarine volcano, are upended here. By examining the pillows closely, geologists can tell which side of the vertical layers was originally up. Can you?

Up the West Fork Trail there are thick sandstone beds deposited by submarine slurries of sand, mud, and water. Such gravity-propelled turbidity currents flow down continental slopes or irregularities in the sea floor. Because sand-mud slurries usually settle and harden as massive, poorly layered sandstone that is coarse near its base and finer near its top, it too can be used to distinguish the original position of the rock.

Elwha River and Rica Canyon. Of all the rivers radiating from the Olympic Mountains, the Elwha is perhaps the most dramatic. It heads nearer the center of the range. A drive up the Elwha will take you through pillow lava of the north arm of the basalt horseshoe into some of the weak, slide-prone sedimentary rocks. In places there are Pleistocene lake sediments that show us that the river was several times dammed by the great ice tongues from Canada.

River-carried cobbles of quartzite and basalt reveal the history of the Olympic Range. Angular dark fragments in the quartzite originated as flakes of mud mixed with sand by undersea currents. Fine veins of quartz developed during metamorphism and uplift.

At Observation Point, where the road turns up Boulder Creek, you can look some distance up the Elwha. The river flows alternately through narrow canyons and wide valleys, depending on the hardness of the rock. At Olympic Hot Springs, just beyond the end of the road, warm water comes to the surface along the Calawah fault zone, one of the prominent faults in these mountains.

Not far up the Elwha River Trail is the narrow defile of Rica Canyon. Its upper end is known as Goblins Gate. There the Elwha, in order to make a right-angle turn into Rica Canyon, churns in a giant whirlpool. All along the Elwha, as far up as the steel footbridge, you can see gravel bars, driftwood, and piled debris left by a 1967 flood caused when the rain-swollen river cut rapidly through a landslide dam just above the footbridge site. Farther upstream the trail climbs above the river to avoid the Grand Canyon of the Elwha, where the river loops through a tortuous, deeply entrenched canyon with hard sandstone walls.

Hoh River. The dense rain forest on the seaward side of the Olympics conceals most of the rocks there. In forest-covered areas, geologists often study the cobbles and pebbles in river beds, secure in the knowledge that they have moved downstream only. How many rock types do you find here? What do they say about the rocks upstream? Several varieties of gray sandstone, some containing fragments of lithified mud, were once the stuff of submarine density currents—slurries of

Fed by glaciers, the Elwha River carries debris from one of many landslides along its course. Compare the angular boulders here with the rounded pebbles of the Hoh River and the Pacific beaches.

sand and mud that flowed like rivers down undersea slopes. Black slate in thin, flat "skipping stones," some with tiny brassy pyrite crystals, suggest that at times the sea muds were black with decaying organic matter. Greenstone, more common as you get farther up the valley, is altered basalt from thin layers interbedded with the sandstones and slates; the upper Hoh nowhere drains any of the main basalt horseshoe. All the rock fragments have been rounded by stream action, bounced along and struck with other rocks until their corners are worn away.

Hurricane Ridge. Even if you are not primarily an automobile explorer, do drive up to Hurricane Ridge. The road climbs to highlands of the basalt horseshoe, with excellent exposures of pillow basalt and volcanic breccia. The volcanic rocks contain little round vesicles, fossil bubbles, now filled with calcite or other light-colored minerals. Bright red rock in some roadcuts is muddy limestone that contains microscopic fossils that helped to date adjacent sedimentary rocks as younger than the volcanic rock of the horseshoe.

Looking southwest from Hurricane Ridge Visitor Center (where there are several geologic exhibits), you can just see Mount Olympus, 2,428 meters (7,965 feet) high, snowcapped, girdled in glaciers, peeking over its nearer neighbors. The dominant peak in front and to the right of it is Mount Carrie, 2,132 meters (6,995 feet) high. Broad Carrie Glacier on its flank is marked late in the summer with deep crevasses. Its terminal moraine and meltwater stream can be seen with binoculars.

Close at hand is the deep valley of the Elwha River, its slopes streaked with the livid scars of avalanche paths and landslides and the light green vegetation of old burns. Landslide, avalanche, slump, and creep are important and normal processes here in the never-ending battle between uplift and erosion.

Most of the heartland of the Olympic Mountains, the terrain you see in the 180-degree view from Hurricane Ridge, is wilderness, just as it was before the coming of the white man. The line of high peaks to the southeast and south, seeming to embrace the rest of the vista, is the resistant volcanic horseshoe. Remnants of an inner, thinner arc of basalt show up as Steeple Rock, Eagle Point, and Obstruction Peak.

Many trails invite the hiker here, and they are fairly easy ones since the road did most of the climbing. Along them can be found outcrops of pillow lava and tilted sandstone and shale layers. Mount Angeles, northwest of Hurricane Ridge, displays vertical ribs of hard sandstone, basalt, and volcanic breccia alternating with thin, soft beds of red shale. The corrugated mountain flank is an excellent example of differential weathering and erosion of hard and soft rock layers. When broken volcanic rocks are deposited in water, large fragments sink to the bottom first and are covered by finer and finer particles. When this coarse-to-fine sequence is used to identify the original position of breccia layers, we see that on Mount Angeles the layers are tilted *beyond* vertical.

Lake Crescent. The long east-west valley now occupied by Lake Crescent and Lake Sullivan was shaped in part by the lobe of Canadian ice that crept across Puget Sound and out the Strait of Juan de Fuca. Some ice squeezed into this valley from the north, and flowed in both directions—east and west—deepening and reshaping stream valleys. Just after the ice retreated, Lake Crescent filled the whole length of the valley and drained out Indian Creek at what is now the east end of Lake Sullivan. A landslide later separated the two lakes, blocking this drainage route for Lake Crescent. Its water level rose until it overflowed northward through the Lyre River channel.

Above the highway along the south side of the lake are exposures of pillow basalt of the north arm of the volcanic horseshoe. Here the hardness of the basalt has superimposed an east-west pattern on otherwise radial drainage.

The Pacific Beaches. The sand and pebble beaches of the western coast show many interesting geologic features. The shore makes inviting walking. (Since headlands are negotiable at low tide only, don't get caught with no way to return.)

The shore in most places is backed by cliffs of stream-deposited glacial outwash sorted as to boulder, pebble, and sand size. Logs and tree trunks flung ashore by storm waves protect the cliffs to some extent, but winter storms continue to alter the shape and width of the beach. Offshore islands are formed of the same rocks as the mainland and were once part of it.

A submerged wave-cut bench edges the coast and reaches out beyond the string of islands. It plays an important part in the molding of this coast, for as waves from the open sea come across the shallow bench, they shorten and steepen, becoming more effective as erosional agents.

Most of the pebbles on the beaches are washed from the glacial gravel of the cliffs, and they represent almost all the harder rock types of the Olympic Mountains. There is no bedrock granite here, but occasionally one finds a granite pebble brought here by the lobes of ice that pushed down from Canada in Pleistocene time.

Point of Arches, with its sea caves, natural arches, and broad, wave-cut bench studded with stacks and other offshore rocks, is typical of the

What are the tales told by the beach? Ripple marks and wavelike mounds of sand describe currents along the shore. Rill marks speak of an ebbing tide. Rounded pebbles (frontispiece) recount a long journey from mountain to sea.

headlands along the Olympic coast. The oldest rocks on the entire peninsula, aside from the granite erratics, are found here: shiny black crystalline gabbro, a coarse-grained equivalent of basalt. This rock is thought to have formed at the Pacific's mid-ocean ridge well before the North American Plate drifted west over it.

At low tide, some of the sea stacks can be reached by wading. Low tide reveals the wave-cut bench from which they rise, as well as the wave-carved notch at the base of both stacks and headlands.

Fist-sized cavities called honeycomb weathering, visible in masses high up on some sandstone cliffs, are etched by wind and spray.

Not far off this coast is an oceanic ridge that marks the west edge of the Juan de Fuca Plate, one of several small crustal plates that stood in the way of North America's westward drift. The continental part of the plate—a microcontinent—has attached itself to North America, with the suture just east of North Cascades National Park.

Queets River. The road up the Queets River goes up a long glaciated valley, cutting through a succession of glacial moraines. The lowest, oldest moraine is almost indistinguishable, so deeply is it weathered. Because of the dense rain forest, the upper moraines are visible only where the river cuts through them. Rapids and several low waterfalls have developed where the Queets encounters hard sandstone layers.

Quinault Lake, Quinault River, and Enchanted Valley. The park boundary hugs the northwest shore of Quinault Lake, a natural lake dammed by the terminal moraine of the largest glacier to come down the valley. Flowing southwest from highlands around Mount Anderson, the river of ice must have been about 50 kilometers (32 miles) long. Glaciers negotiate corners with difficulty; usually they straighten the river valley down which they flow, as this one has done, by truncating the spurs that edge it. Hanging tributary valleys show that small tributary glaciers were unable to cut down as fast as the main one.

Many geologic details are hidden by thick forests here, and by talus and rock piles that accumulated where valley walls collapsed after retreat of the glaciers. Landslide and creep still work to reduce those walls.

Enchanted Valley, at the head of the East Fork, is a picture-book alpine valley, deeply U-shaped and with cliffs of resistant sandstone. The small moraine at its lower end, and the extremely flat floor of the valley, tell us that there was once a moraine-dammed lake here.

Seven Lakes Basin. On the north-facing slope of Bogachiel Peak, little glacial lakes called tarns lie in glacier-scoured depressions in the floor of a large cirque. This is the birthplace of a glacier that combined with the Soleduck Glacier to scour Soleduck Valley. The easiest access to this basin is by trail up the Soleduck River, though it can also be reached from the west by much longer trails. Glacial features here have been modified to some extent by collapse of steep glacial cliffs, since the ice that both carved and supported them melted away. Notice the talus cones here, and the scars of landslides. The blocky sandstone of Bogachiel Peak, easily broken along numerous joints, continues to slide and creep downhill. Some sandstone layers display graded bedding, the coarse-to-fine sequence that indicates the original position of steeply tilted strata.

Soleduck Hot Springs. The hot springs and spa here owe their existence to much-fractured sandstone through which water heated deep down in the crust can rise rapidly. The depth of Soleduck Valley shortens the distance it must flow to reach the surface. The water is chemically like surface water and probably seeps down from rivers and streams to a heat-exchange region where it comes in contact with hot rocks. The change in density as it heats causes it to rise to the surface again.

At nearby Soleduck Falls, the river tumbles through a deep cleft etched into a vertical bed of soft shale. The falls are bordered by more resistant sandstone layers.

OTHER READING

Danner, W. R. 1955. *Geology of Olympic National Park.* University of Washington Press, Seattle, Washington.

Rau, W. R. 1973. *Geology of the Washington Coast between Point Grenville and the Hoh River.* Washington Department of Natural Resources, Geology and Earth Resources Division Bulletin 66.

Sharp, R. P. 1960. *Glaciers.* University of Oregon Books, Eugene, Oregon.

Snavely, P. D., Jr., and Wagner, H. C. 1963. *Tertiary Geologic History of Western Oregon and Washington.* Washington Division of Mines and Geology Report of Investigations 22.

Tabor, R. W. 1965. *Geologic Guide to the Deer Park Area, Olympic National Park.* Olympic Natural History Association, Port Angeles, Washington.

Tabor, R. W. 1969. *Geologic Guide to the Hurricane Ridge Area.* Olympic Natural History Association, Port Angeles, Washington.

Tabor, R. W. 1975. *Guide to the Geology of Olympic National Park.* University of Washington Press, Seattle, Washington.

Oregon Caves National Monument

Established: 1909
Size: 1.9 square kilometers (0.74 square mile)
Elevation: 1,226 meters (4,020 feet) at cave entrance
Address: 19000 Caves Highway, Oregon 97523

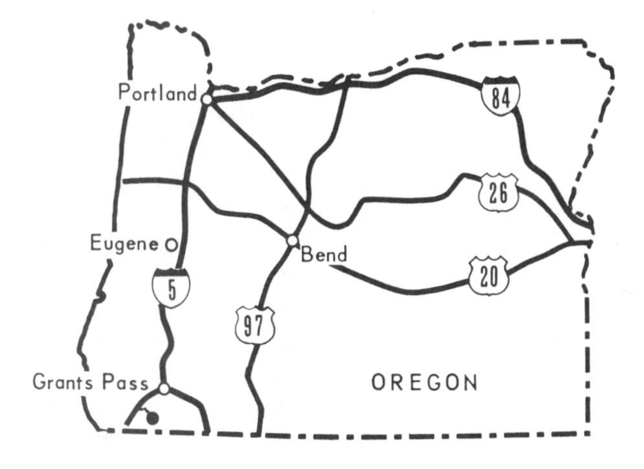

STAR FEATURES

• A cavern dissolved in steeply tilted Mesozoic marble, ornamented with a variety of flowstone and dripstone features.

• An underground stream, the River Styx, called Cave Creek once it leaves the cavern.

• Guided cave tours led by Park-Service-trained guides, a self-guiding nature trail, and other trails. Visitor center facilites are minimal.

SETTING THE STAGE

In the heart of the Siskiyou Range, this cave (there is just one) occurs in a region of folded and faulted sedimentary, metamorphic, and volcanic rocks. The bluish marble in which the cave formed is steeply inclined, with its upper, western edge coming to the surface along the west side of Mount Elijah.

As much as 120 meters (400 feet) thick in places, the marble, altered by heat and pressure, originated as marine limestone. It is not fine, decorative marble, but contains impurities such as clay and silt and derivatives of volcanic rock. Like the limestone from which it developed, the marble is made up of the mineral calcite, and like calcite it is soluble in

Oregon State Highway Dept. photo

In one of Oregon Cave's passages, stalactites and stalagmites merge into massive columns. Original layering of the marble is apparent in the wall rock.

dilute acid. Therein lies the clue to the first step in cave construction—solution. Rainwater, picking up carbon dioxide from the air, is mildly acidic and becomes more so as it filters through soil to become groundwater.

If you examine the marble near the cave entrance or along the road leading to it, you'll see that it is cracked and jointed, a result of uplift and folding. The joints, many of them parallel, were also important to cave formation, for it was via paper-thin cracks and crevices such as these that groundwater first traveled through this rock. Seeping slowly, it gradually dissolved away a little of the crevice walls, enlarging a few favored passages. Wider passageways of course allowed faster flow, which in turn sped up the dissolving process and enlarged the passages even further.

Most cave solution takes place just below the water table, where the rock is completely saturated, with all its crevices and pore spaces completely filled with water. Even deep underground, the water still responds to the tug of gravity and flows through the rock toward springs and streams, ponds, lakes, and ultimately the sea. Usually its underground flow is roughly parallel to the surface of the ground. Here the water table slopes more gently than the ground above it, approaching the surface right at the cave entrance.

The water table has now dropped below the level of most of the cave. What accounts for this change? Deepening of downstream canyons no doubt draws off much of the water. New and unknown features of the underground plumbing system may have pulled the plug, so to speak, allowing the water to drain away somewhere below the surface. Within the cave there is evidence that the water table fluctuated at times. Solution is always most rapid near the surface of the water table, so we can guess that lines of deep grooving and pitting represent old water levels, where the water table remained stationary for a time. The level now fluctuates a little with the season; the cave becomes perceptibly drier during the almost rainless Oregon summer.

Lowering of the water table brought about the second step in cave formation—ornamentation. Most types of cave ornaments form in air. Dripstone and flowstone develop where water loaded with dissolved calcium bicarbonate is exposed to air—

Oregon State Highway Dept. photo

Flowstone draperies of the Paradise Lost area hang like crowded crystal chandeliers. These features continue to form as lime-laden ground-water continues to lose carbon dioxide to the cave air.

D. D. Givens photo, courtesy N.P.S.

Pure and white, a new-formed flowstone "waterfall" tells the story of its growth. Lime-laden water seeps from a crack in banded marble of the cave wall, trickles to a ledge, spreads out, and flows in separate streams across a massive earlier incrustation.

even to the moist air within the cave. Both are made of calcium carbonate, and both owe their development to the same process. On reaching the air-filled cavern, water seeping down from above loses some of its carbon dioxide to the air. With less carbon dioxide, the water cannot hold as much calcium carbonate, so it leaves behind a thin coating of calcite. Through thousands of years, as water drips and flows from narrow cracks and joints, these minuscule deposits add up into strange and spectacular speleothems, cave ornaments.

The most common speleothems are stalactites, stony icicles that project downward from the cave ceiling. Below many stalactites are sturdy, round-topped pillars called stalagmites. (One way to remember which is which is to keep in mind that stalactite is spelled with a "c" for ceiling, stalagmite with a "g" for ground.) Water dripping from the ceiling creates the slender stalactites; water splashing onto the surface below forms thicker stalagmites. Both are built of concentric layers of radiating calcite crystals that continue to grow at their outer ends as long as lime-bearing water is flowing over them. Where stalactites and stalagmites join, they form columns.

Draperies and "frozen waterfalls" form as heavily mineralized water flows over ledges in the walls of the cave. These flowstone ornaments are often decoratively banded with impurities. In them, calcite crystals tend to be at right angles to the flow surface. Both flowstone and dripstone are often known as "cave onyx," a translucent, banded form of calcite.

This cave contains some very delicate, hollow soda-straw stalactites that form as lime-laden droplets deposit fine rings around themselves. Successive rings extend the hollow tube. Some soda straws are astonishingly long; many are so delicate that you can see through them.

Cave "popcorn" or globularites also occur here. Unlike other cave ornaments, they are believed to form under water, particularly near the surface of still water. When they occur on stalactites and stalagmites, we know that a cave once above the water table must for a time have been resubmerged.

GEOLOGIC HISTORY

Mesozoic Era. About 200 million years ago, in Triassic time, this region was below the sea—a replay of earlier submergences during the Paleozoic Era. Shells of marine denizens of various types—clams and snails, corals, and microscopic one-celled protozoa—accumulated on the bottom of the shallow sea as a limy ooze that was to become the Applegate Formation. Eventually other sediments like sand, clay, and gravel covered the limestone. Streams of lava from nearby volcanic islands spread out on the sea floor, too, from time to time. The scene may have been almost a mirror image of the area around Japan today, where an arc of volcanic islands is separated from the mainland by a shallow sea in which marine sedimentary deposits are interlayered with volcanic flows.

Starting possibly in Jurassic time, the region became the prow of North America's westward drift, if a raft as big as North America can be said to have a prow. Carried west, the leading edge of the continent pressed hard against the edge of the East Pacific Plate. Tremendous pressures exerted on the continent as it moved up and over the East Pacific Plate caused folding, crumpling, and eventual metamorphism of the old Triassic sea-floor sediments, among them the Applegate Limestone. In the process, the rocks were steeply tilted, compressed, folded, and probably heated at depth. They ended up as newer, harder rock—marble. All traces of the shells of which they were made were erased as minerals in the rock recrystallized, so that now their age can be determined only by studying adjacent sedimentary and volcanic rocks.

Cenozoic Era. As the region rose it was of course attacked by erosion. Slowly the sandstone and shale and lava layers above the marble were stripped back, so that gradually the marble became less and less deeply buried. Finally it was within range of acid-charged water seeping down from the surface, and cave solution began. Solution no doubt benefited from climatic changes, especially those of Pleistocene time when periods of high precipitation brought ice caps to nearby mountains and heavy rains to this and other lowland areas. Then, as the Ice Ages waned, the cave became drier and growth of cave ornaments began. Possibly one or more fluctuating repeats of Ice Age conditions brought resubmergence, allowing popcorn to form on stalactites and stalagmites. In any case, final lowering of the water table initiated more flowstone and dripstone growth, making the cave what it is today.

Construction of dripstone and flowstone ornaments is still going on here. Some tiny stalactites have developed near the man-made entrance, for example. The River Styx is deepening its channel. And who knows what further channeling is taking place below the present water table and what other caves are hidden in these hills?

OTHER READING

Contor, R. J. 1963. *The Underworld of Oregon Caves National Monument.* Crater Lake Natural History Association.

Pinnacles National Monument

Established: 1908
Size: 65 square kilometers (25 square miles)
Elevation: 305 to 1002 meters (1000 to 3287 feet)
Address: Paicines, California 95043

STAR FEATURES

• The remains of an ancient stratovolcano, whose explosive eruptions 23 million years ago spread layer upon layer of volcanic rocks—lava flows, tuff, and breccia—around its several vents.

• Remnants of one or more volcanic domes formed as thick, viscous lava squeezed up out of feeder dikes.

• Pinnacles, crags, and balanced rocks carved by water and wind in volcanic ash and breccia; interesting caves formed in a most unusual manner: by falling rocks.

• Evidence of an earlier route of the San Andreas Fault, and a measure of displacement along it.

• Visitor center, and many trails (some, including a geologic trail, with guide leaflets) from which geologic features can be seen close at hand.

SETTING THE STAGE

From the rugged summits and steep slopes of the southern Gavilan Range rise the pinnacles, domes, and spires of volcanic tuff that give this monument its name. The tuff occurs in confined, sharply outlined, oval-shaped areas 500 to 1500 meters (0.3 to 1 mile) across, as well as in more widespread layers between thick sloping sheets of volcanic breccia, product of explosive volcanism.

Pointed spires of rhyolite tuff rise above the trail, their surfaces case-hardened by weathering processes. Chalone Peak (background, right) is a remnant of a lava dome.

Vincent Matthews III photo

The primary volcanic rock type here is rhyolite, a light-colored (usually pinkish gray) silica-rich volcanic rock. Rhyolite occurs as tuff, flow-banded lava, volcanic breccia, and volcanic glass. Several of these variations may form during the same eruption: the tuff from volcanic ash, the flow-banded lava of course from lava flows, volcanic glass where molten lava cools particularly quickly, as at the frothy top of a flow or where it contacts water or a chilly underlying surface. Volcanic breccia forms where large rock fragments thrown from a volcano's vent fall to earth. Tuff in oval outcrops is thought to be ash that filled volcanic vents as eruptions died down; five such vents occur in the national monument. Rhyolite boulders, rounded and well mixed with finer material, appear in what seem to be mudflows.

Some of the rhyolite is layered, with west-tilting layers probably close to their original position on the flanks of a growing volcanic cone. Elsewhere it is massive, with no trace of layering. When it is too thick and pasty to flow, rhyolite magma may squeeze up out of fissures like toothpaste from a giant tube, to form lava domes.

Erosion of the spires and pinnacles—the work of water and wind—followed numerous vertical joints that served as avenues for seeping water. Weathering processes tend to harden volcanic tuff, cementing together the near-surface grains. Once the surface crust breaks through, though, wind and water may hollow out small alcoves. Larger caves within the monument occur where huge blocks of massive rhyolite flows or breccia have fallen into narrow ravines, to wedge between rock walls. The entire volcanic area is about 38 kilometers (24 miles) long from north to south, giving us a rough estimate of the size of the original volcano. To east and west, the volcanic rocks have been removed by motion along two prominent faults.

The faults that border the volcanic area—the

Vincent Matthews III photo

Massive layers of tuff and breccia protrude from hills along the Condor Gulch Trail. Some contain tiny fossils, showing that some of the tuff was water-laid.

Breccias consist of highly fragmented volcanic rock thrown from a volcano's vent. These rounded boulders, imbedded in volcanic debris of many sizes, suggest a mudflow origin.

Vincent Matthews III photo

Chalone Creek Fault on the east, the Pinnacles Fault on the west—trend about north-northwest. They are related to the San Andreas Fault just a few kilometers east of the national monument, which trends in the same direction. The Chalone Creek Fault, probably still active, is thought to mark the position of the San Andreas Fault back in Miocene time—at the time the Pinnacles volcano erupted. So the Pinnacles area is part of the western sliver of California that is moving northward in small increments because of horizontal displacement along the San Andreas Fault.

Moreover, the Pinnacles area is a major clue to movement on the San Andreas Fault. Volcanic rocks of the Pinnacles Formation, all west of the Chalone Fault, are dead ringers for some volcanic rocks near Los Angeles, just east of the San Andreas Fault—identical rock types in an identical sequence. Probably once connected, they have been offset—pulled northward—315 kilometers (196 miles) along the San Andreas Fault, at an average rate of 1.4 centimeters (roughly half an inch) per year since Miocene time.

GEOLOGIC HISTORY

The oldest rocks in this area are metasedimentary rocks of the Sur Series—marble, slate, and quartzite probably of late Paleozoic (Pennsylvanian or Permian) age. These rocks occur in scattered outcrops outside the national monument, embedded in the granite that engulfed them in Cretaceous time. Both the metasedimentary rocks and the granite apparently underwent a long period of erosion in early Tertiary time—erosion that wore away almost all of the metasedimentary rocks and "unroofed" the granite before eruption of the Pinnacles volcano.

About 23 or 24 million years ago, during the Miocene Epoch (a time of volcanism in much of western United States), rhyolite magma forced its way up through cracks in the crust. Some of the magma broke through to the surface, erupting first as fine volcanic ash, then as flowing lava which was to become the flow-banded rhyolite of the eastern part of the monument. The volcanic ash that records the beginning of volcanic activity is now a thin layer of glassy tuff. At the top of the flow-banded rhyolite is more volcanic glass, in the form of the beady perlite now visible near monument headquarters.

Later, after several andesite and dacite lava flows, another type of eruption occurred: Thick, pasty lava squeezed upward and outward in glowing, dark-crusted lava domes. The massive rock of these domes now makes up the eastern slopes of North and South Chalone Peaks, in the southern part of the national monument. The steep-sided domes became surrounded with talus and rockslides, much as Lassen Peak, a more recent volcanic dome in Lassen Volcanic National Park, is today surrounded by a talus skirt.

After a period of quiescence, new eruptions began, more violent ones. Great explosions shot out broken rock, molten blobs of lava, and hot volcanic ash. With repeated eruptions, the older flows and the massive domes were covered over. As explosive eruptions alternated with quiet venting of lava flows or fine volcanic ash, a stratovolcano took shape. Ultimately it would measure at least 40 kilometers (25 miles) across at its base, and perhaps 2500 meters (about 8000 feet) high—perhaps a graceful conical mountain not quite as large as Mount St. Helens. Debris from the explosive eruptions makes up the band of volcanic breccia in the western part of the national monument, the region of the highest peaks.

Given the barren slopes of the growing volcano to work on, the forces of erosion cut gullies and gulches, and transported boulders and pebbles of breccia, as well as quantities of volcanic ash, down the mountain slopes. Remains of this water-carried debris, in the northeastern part of the national monument, can be recognized by their stream-rounded boulders and cobbles.

Changes in land level frequently changed the patterns of deposition and erosion. The volcano at times had its feet in the water, as is shown by a now prominent layer of white diatomite made of the silica skeletons of innumerable tiny aquatic plants, and by the shells of marine organisms found in some of the tuff. Several lines of evidence suggest that the massive breccia of the western part of the monument was deposited under water. In several places alluvial fan and mudslide deposits grade into nearshore types of marine sedimentary rock.

Movement along two approximately parallel faults—the Chalone Creek Fault on the east and the Pinnacles Fault on the west—saved the volcano from being worn away entirely. By chance, or because of the collapse of a partly depleted magma chamber well below the surface, a long north-south sliver—the region of the present national monument—sank several hundred feet, while adjacent areas rose. Protected for a time from erosion, the sunken block remained more or less intact, though higher parts of the mountain were destroyed. The central sliver includes five vents, a long north-south strip of rhyolite lava, and on the west side some of the layered volcanic breccia and tuff of the volcano's western flank.

The rest of the story is one of fault movement and erosion. Since Miocene time, movement along the San Andreas Fault carried the sunken block bearing the remains of the Pinnacles volcano far north of its original southern California position. As it moved northward, rain and running water cut steep, narrow canyons, such as those of Bear Creek and Chalone Creek, into the volcanic rocks. Patches of tuff, particularly the tuff that had fallen back, as volcanic ash, into the several volcanic vents, were attacked along closely-spaced vertical joints, so that pinnacles developed. In places erosion cut into thin tuff layers, undermining blocks of the more resistant breccia, forming mushroom rocks. Elsewhere, huge masses of undermined breccia slid, slumped, or tumbled down steep canyon slopes, wedging themselves into narrow ravines, forming the roofs of today's caves and tunnels.

OTHER READING

Webb, Ralph C. 1969. *Natural History of the Pinnacles National Monument.* Pinnacles Natural History Association.

Redwood National Park

Established: 1968
Size: 442 square kilometers (171 square miles)
Elevation: Sea level to 944 meters (3,097 feet)
Address: 1111 Second Street, Crescent City, California 95531

STAR FEATURES

• A section of the Coast Range, with geologically young sedimentary and volcanic rocks scraped from the sea floor and crumpled into mountains.

• Shore features created by coastal processes: wave-carved cliffs, sea stacks and caves, sand and pebble beaches.

• Tall coast redwoods, "living fossils" that survive from vaster forests of the past.

• Evidence of geologic events resulting from clear-cutting practices of the last 50 years.

• Visitor center exhibits, evening programs, guided and self-guided tours, scenic drives and trails (some with wayside displays). The interpretive program is evolving rapidly with changes in park boundaries.

See color pages for additional photographs.

SETTING THE STAGE

In California's northwest corner, the Coast Range narrows to a slender band of steep-sided ridges and deep valleys not quite parallel to the coast. On their western side, cliffs plunge sharply into the sea; in places there are narrow strips of beach. The continental shelf, its shoreward edge beveled by pounding waves, reaches westward from the coast. One major river, the Klamath, crosses the park from the southeast. Climatically the region is one of cool summers, rainy winters, and frequent fogs, favoring growth of conifer forests dominated by stately redwood groves.

The geology of the Coast Range is as complex here as it is elsewhere. Though rocks inland from the coast are hidden by vegetation and therefore difficult to study, there are good exposures in the coastal cliffs. Most of the rocks belong to the Franciscan Series, a catch-all name for a group of drab brown and gray sandstone, siltstone, and conglomerate altered enough to be called metasediments, interspersed with schists and patches of chert and altered submarine basalt called by the descriptive

In 1964, Redwood Creek floodwaters threatened to undermine this cluster of tall trees, among them the world's tallest (second from left). Recently improved land-use practices and erosion control in clearcut areas upstream may lessen the impact of future rainy-season floods.

Franciscan Series slate and sandstone, offset by vertical faults, are exposed at roadside near the Klamath River bridge.

but somewhat imprecise term greenstone. Much of the sandstone is graywacke, in which individual sand grains are not individual mineral crystals but small fragments of pre-existing fine-grained rocks such as shale and mudstone. The graywacke is thought to result from rapid erosion of a rising coastline, plus quick deposition by turbidity currents moving rapidly down undersea slopes. At the north end of the park there is a little serpentine—a streaky greenish rock—and some black gabbro, rocks that may have seen their beginnings in the Earth's mantle below the oceanic crust.

The occurrence together of sedimentary and volcanic rocks, the one derived from the continent, the other oceanic, provides a clue to the mechanics involved in creation of the Coast Range—mechanics now explained in terms of the Plate Tectonic Theory. As the North American Plate pushed west, overriding the East Pacific Plate, like a bulldozer it scraped the surface over which it rode, scooping up layers of mud and sand as well as some of the volcanic rocks of the sea floor. Crumpled and jumbled together, the oceanic lava and continent-derived sediments were badly contorted as if they were scraped up and broken and bent before they were fully lithified, when they were as pliable perhaps as mudpies. Some, pulled down for a time into a deep oceanic trench along the continent's margin, were baked and compressed to such an extent that they recrystallized into silvery, mica-rich phyllite or, in the case of sandstone, into quartzite. Rarely, though, are they as completely remelted or recrystallized as the granite and gneiss of California's interior mountains.

Despite poor exposures, it is easy to find good examples of these rocks. Look on the beaches and in stream beds, where pebbles and cobbles abound, green and tan, silver and dull gray. They are also exposed in sea cliffs, in new roadcuts, and in some river banks.

Here as elsewhere, geologic processes shape the land. Relentless rains gather in rivulets to wash away bare soil wherever it is exposed. Little by little, soil creep feeds materials of the mountain ridges to the streams below, which relay them to rivers and thence to the sea. The coastline changes as it is pounded by waves and swept by currents. Beaches widen where rivers replenish their sand, or narrow when this material is denied them. Coastal landforms are varied: cliffs and headlands, estuaries, beaches, quiet lagoons. Isolated remnants of an older coastline stand as sea stacks beyond the line of surf. Waves surging against headlands carve caves and arches and level the wave-cut bench that stretches seaward beyond the shore. (For more on the origin of these coastal landforms, see Olympic National Park.)

Man and his works feature in the story of erosion here, more so than in most national parks. Ruthless timbering has invited a speedup in erosional processes over entire drainage basins. Winter rains, unhindered by vegetation, deepen furrows left by dragged logs and heavy logging equipment and carry away the loose soil once held by plant roots. Winter rain, no longer absorbed by a spongy humus layer, swells streams that are already overloaded with rocks and sand. Rampaging rivers tear at aged redwood groves. For this reason 194 square kilometers (75 square miles) of private timberland were added to the park in 1978, and 120 square kilometers (47 square miles) of land adjacent to the southern part of the park, much of it already logged, has now been declared a park protection zone. In this zone erosion is being slowed by any means possible, and vegetation is being re-established in an effort to protect and preserve the groves of exceptionally tall trees downstream in the park.

GEOLOGIC HISTORY

Mesozoic Era. Sedimentary rocks of the Franciscan Series were deposited along the west edge of the continent about 150 to 100 million years ago, during parts of the Jurassic and Cretaceous Periods, when North America was still one with Europe. Streams and rivers that then drained the western part of the supercontinent flowed with gentle gradients across a shelving coastal plain similar to that of the southern United States today. Sediments delivered by the rivers to the sea settled as mud, silt, and sand on the shallow continental shelf or flowed down the continental slope in turbid currents to lodge in deeper water on the ocean floor.

Roughly 100 million years ago, the picture began to change. Europe and America parted ways; between them the Atlantic Basin widened. As the North American Plate drifted westward, it plowed up considerable quantities of sea-floor mud and volcanic rock, and possibly some of the underlying mantle, as we have seen.

Cenozoic Era. The North American Plate continued to forge westward in Cenozoic time, and still does so today at a pace of a few centimeters each year. Eventually the crumpled sediments, compressed for a time in an oceanic trench, were thrust above the sea to form the Coast Range.

Coast Range topography reflects the structure of these plowed-up rocks. Northwest-trending ridges and valleys parallel northwest-trending strips of sandstone and schist and volcanic rocks, some quite resistant, some readily eroded. The two major faults in the park—one along Redwood Creek and the other at the north end of the park—show this trend as well, as do the courses of the West Fork of the Smith River, the Klamath River inland from its mouth, and Redwood and other creeks.

Geologic history is still in the making here, on a measurable scale. The coast is rising, parts of it about 3 millimeters (0.1 inch) a year, one of the most rapid uplift rates along this side of the continent. The flat, wave-cut terrace at Point St. George, north of Crescent City and outside the park, formed at tide level at a time when the coast was lower. Pounded by waves, cliffs recede and beaches are created, swept away, and re-created. With gradual uplift, rivers change course.

In Pleistocene time the Klamath River approached the coast about 12 kilometers (8 miles) south of its present mouth. There it built a thick sand and gravel delta. The old delta sediments can be seen now in Gold Bluffs, above Gold Bluffs Beach.

The coast redwoods are part of the geologic story here, too. Redwoods first appear in the fossil record in Cretaceous rocks about 100 million years old. By early and middle Tertiary time, 65 to 12 million years ago, redwood forests had spread across Europe, Asia, and much of North America. Fossil trunks, cones, and foliage show that at least twelve species had evolved. But uplift of continental interiors in late Tertiary time spelled cooler, drier climates, and disaster for the redwood clan. Only three species now remain: the giant sequoia of the Sierra Nevada, the coast redwoods along the California and Oregon coasts, and the dawn redwood, an archaic species that has managed to cling to life in a remote part of China, where only about 1,000 trees remain. The redwoods probably owe their 100-million-year survival to an unusual degree of built-in resistance to insects, fungi, and fire. Unfortunately they have no way to resist man's greed, and the rich forests that extended along this coast and along the Sierra less than 100 years ago are now whittled down to isolated groves. Living coast redwoods are protected here, in John Muir National Monument, and in several California state parks. Giant sequoias grow in Sequoia, Kings Canyon, and Yosemite National Parks. Florissant Fossil Beds National Monument in Colorado preserves a group of fossil sequoia trunks; others occur in Yellowstone National Park.

BEHIND THE SCENES

Bald Hills Road. Climbing one of the many parallel ridgelines of the Coast Range, this road heads into country once covered with luxurious stands of redwoods and Douglas firs. Parts of the road traverse heavily logged segments now included in the expanded park—a good place to see the destruction of forest, undergrowth, and soil wrought by past logging practices. Geologic events resulting from clearcutting include landslides, gullying, accelerated soil creep, and floods. The summer shuttle bus to the Tall Trees Trailhead will take you part way up the road; a park naturalist usually rides the shuttle to explain the battle against destruction of the remaining redwoods along Redwood Creek.

Coastal Trail and the mouth of the Klamath. High above the sea, this windswept trail and the shorter nature trails at both trailheads give abundant views of the sea cliffs, offshore sea stacks, and beaches of this stretch of the coast, with trailside displays explaining some coastal processes. Waves strike the shore at an angle here, setting up a southward current that continuously straightens the beaches and carries away debris washed from the cliffs. North of the Klamath River, beaches are skimpy, as you can see. South of the river, where they are amply supplied with river sand carried south by the longshore current, beaches are better developed.

The current reworks sand brought by the river to its mouth, developing a spit that nearly closes the channel. If the tide is ebbing, you can see the contact between fresh water and salt, out beyond the spit. (It may come as a surprise that they don't mix immediately.) Gulls often hover near this line to feed on freshwater fish and crustaceans brought down by the river. When the tide is in, salt water extends into the estuary.

Longshore and tidal currents vie with the Klamath River for mastery of the river mouth, shaping the gravel spits that nearly close the estuary.

Fern Canyon. This lovely glen, its vertical walls draped with ferns, was carved by Home Creek as it cut down through uplifted gravel deposits of the former Klamath River delta. (For the story of this delta, see Gold Bluffs Beach, below.) Fern Creek's drainage area is small, so it now babbles lightly through a canyon much too large for it, over pebbles of metamorphic and volcanic rock types unknown in the Coast Range, brought a longer distance by a larger river. The same pebbles are visible in the delta deposits of the canyon walls and the Gold Bluffs.

Gold Bluffs Beach. The bluffs here are quite different from the dark, somber cliffs of Franciscan rocks found elsewhere along this coast. Their horizontal layers of sand and gravel are known as the Gold Bluffs Formation and were deposited about 2 million years ago by the Klamath River, which at that time reached the sea here, south of its present mouth. Some of these delta deposits display festoons of cross-bedding characteristic of river and stream deposits. Others contain carbonized wood, fossil pine cones, and clam and snail shells (protected of course by national park rules against fossil collecting). The cliffs are stained yellow with rustlike iron oxide, the mineral limonite, developed from decomposing iron minerals disseminated through the sandstone and conglomerate.

The beach here was virtually nonexistent in the 1850s. Then, enterprising miners waited for low tide to mine gold from delta sediments washed and winnowed by surging high-tide waves. They dug where the waves had sorted out black sand, which is made up of heavy iron-bearing minerals and even heavier gold, and then washed the black sand in rockers and sluice boxes. The source of the gold is in the Klamath Mountains farther inland, mountains that like the northern Sierra contain igneous and metamorphic rocks enriched with gold-bearing veins.

The shape of the present beach and the development of soil and vegetation below the cliffs indicates that there has been a string of lagoons behind the beach dunes. Most of the beach sand comes from the present Klamath River mouth; it is brought here by the longshore current. Growth of the beach since mining days can be accredited to mining and logging practices along the upper Klamath River. Hydraulic mining (now outlawed), in which gold-bearing gravels were washed with powerful fire-hose streams of water, flushed huge quantities of sediment into the river and its tributaries. Simultaneous destruction of vegetation brought about increased flooding, speeding up the river processes that bring sand and gravel to the coast.

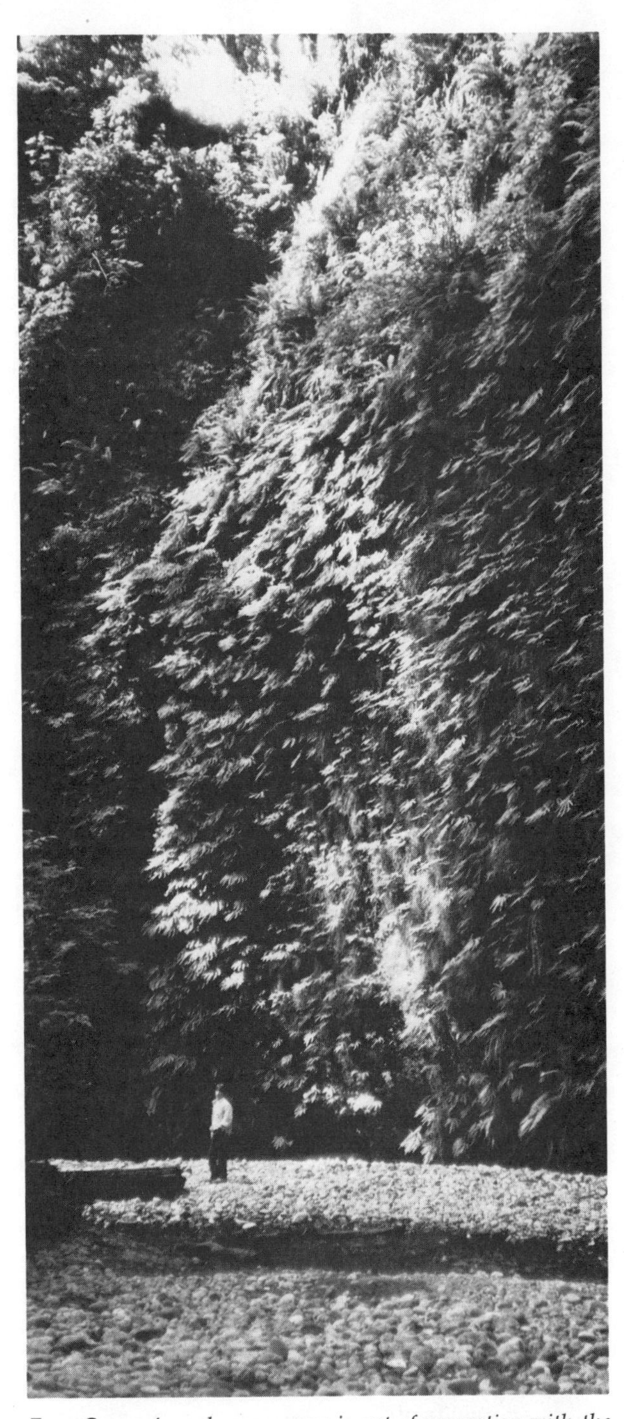

Fern Canyon's cool green gorge is out of proportion with the present stream. Its pebbly floor reveals gravel from the Klamath Range.

Notice the slumps and landslides in the bluffs, and the talus piles of sand and pebble rubble at their base. Though 2 million years have passed, the old Klamath delta deposits have not yet become really well consolidated. Before the growth of the beach, when the sea surged right up to the bluffs, the bluffs probably receded faster than they do today. There is no way to tell exactly how much farther out to sea the coastline lay 2 million years ago when the Klamath had its delta here, but certainly it must have been several kilometers farther west.

Klamath River. Its gravels rich in gold, the Klamath was the scene in the last century of a gold rush of its own. The river is born in Klamath Marsh in south central Oregon, and flows across northern California and through both the Klamath Mountains and the Coast Range. Its cobbles and pebbles reflect its drainage area: They contain types of rocks and minerals (including gold) not represented in the Coast Range.

Most of the cobble-covered bars and terraces in and near the river resulted from a 1964 flood, which destroyed the historic old town of Klamath and swept away a large bridge that crossed the river some distance below the present bridge. During the summer months, when river discharge is low, the spit at the river mouth grows in size. In winter, which is the rainy season here, the river's volume increases and most of the spit is washed away.

Redwood Highway (U.S. 101). In its coastal portions the highway offers views of coast and sea, cliff-edged headlands, offshore sea stacks, arches, beaches, and other features of this rising, eroding coastline. In forested areas vegetation hides the geology; even roadcuts are covered with moss and other plants that enjoy rainy winters and foggy summers. However, some relatively new highway cuts near the Klamath River, along the stretch of highway rebuilt after the 1964 flood, reveal the nature of much of the Franciscan rock of the Coast Range.

A lily-padded lagoon near the highway just south of the Klamath bridge occupies the channel of one of the river's tributaries. Freshwater Lagoon at the south end of the park is a former bay isolated from the sea by a straight coastal bar, the product of currents that sweep sand south along the coast. Be they stable or sinking or rising, coasts tend to be straightened by the sea, which carves away headlands and seals off bays. Lagoons eventually fill in with plant matter and sediment.

Other features along this highway are discussed separately above.

Smith River. Crossing the north end of the park, the Smith River cuts into Franciscan sandstone, siltstone, and greenstone. West of Hiouchi near the highway bridge, both river and highway cross a major fault that brings to the surface some lustrous serpentine veined with quartz, as well as some dark, massive gabbro. Both are thought to have originated in the Earth's semifluid mantle below the solid crust. They too were caught up by the leading edge of the continent as it bulldozed its way westward, giving us a rare sample of the Earth's makeup underneath the crust.

Tall Trees Trail and Redwood Creek Trail. Besides the contrasts offered by logged and unlogged land, and besides the thrill of seeing several of the world's tallest trees, the downhill hike from Tall Trees Overlook offers interesting glimpses of river processes augmented to flood stage by upstream clearcutting. Most of the pebbled bars and terraces developed in 1964, when flood waters narrowly missed the tallest trees and brought to the attention of conservationists and park administrators the importance of preventing further logging in this drainage basin.

The trail along Redwood Creek follows an old logging artery, with new growth covering most geologic features. The creek parallels a major northwest-southeast fault that separates layered sandstone on the right bank from phyllite on the left bank. Both belong to the Franciscan Series. In several places along the trail you'll see the shiny, silvery phyllite; look for it also in cobbles of the river bed. Flat, parallel mica crystals give it its sheen.

OTHER READING

Bailey, E.H. 1966. *Geology of Northern California.* California Division of Mines and Geology Bulletin 190.

Bailey, E.H.; Irwin, W.R.; and Jones, D.L. 1964. *Franciscan and Related Rocks and their Significance in the Geology of Western California.* California Division of Mines and Geology Bulletin 183.

Sequoia and Kings Canyon National Parks

Established: Sequoia 1890; Kings Canyon 1940
Size: Sequoia 1631 square kilometers (630 square miles); Kings Canyon 1862 square kilometers (719 square miles)
Elevation: Sequoia 457 to 4418 meters (1500 to 14,494 feet); Kings Canyon 1189 to 4341 meters (3900 to 14,242 feet)
Address: Three Rivers, California 93271

STAR FEATURES

• "Living fossils"—giant sequoia trees, one of only three living species of a once-abundant family nearly extinguished by the spread of glaciers in Pleistocene time.

• Deep river-carved canyons, glaciated in their upper regions, dissecting the gentle west slope of the Sierra Nevada. Kings Canyon is one of the deepest in America.

• High country—the double crest of the southern Sierra Nevada—scoured by glaciers and sharpened by freeze-and-thaw weathering. Mount Whitney is the highest point in the "lower 48" states.

• The Sierra Nevada Batholith, with many of its individual granite intrusions (plutons) well exposed in the canyons and high country.

• Roof pendants of metamorphic rocks that hang down between plutons of the great Batholith.

• Prominent granite domes resulting from pressure-release joints in massive, relatively unfractured granite.

• Visitor centers, roadside and trailside displays, daytime and evening programs, guided hikes, and an abundance of trails (many with guide leaflets). Topographic maps are available at visitor centers.

See color pages for additional photographs.

SETTING THE STAGE

The Sierra Nevada, nearly 700 kilometers (430 miles) long and 65 to 130 kilometers (40-80 miles) wide, is the largest fault block range in the country. Most of it is a single huge block of the Earth's crust, a giant west-tilted block broken off abruptly on its upraised eastern edge. The gentle western slope of the block slants like a ramp down to and beneath the thousands of meters of sediments that fill in California's Central Valley. The range is made up primarily of intrusive igneous rocks, granite and its kin, in many individual intrusions, or plutons, together making one giant Batholith.

Pale gray Sierra granite is cut by pressure-release joints concentric with exposed surfaces.

Some quite large and others much smaller, the plutons are composed of coarse-grained igneous intrusive rocks in which you can see the individual mineral grains without magnification. The plutons differ from one another in color, grain size, mineral make-up, and inclusions—fragments of darker or lighter rock imbedded within the main rock. Rock types range from very dark gabbro to dark gray diorite, light gray granite and granodiorite, and almost white alaskite. The color and classification depend on the amount of dark minerals such as hornblende or biotite (black mica), as well as on the relative quantities of quartz and the types of feldspar. All the light-colored, coarse-grained intrusive rocks can be referred to as "granitic rocks," or even, in its broadest sense, as "granite"—a usage employed here.

Composed of chunky grains of quartz and feldspar, granite is peppered with black mica (biotite) and rodlike crystals of hornblende.

There are metamorphic rocks within the Sierra Nevada as well—dikes and irregular masses of tightly folded, complexly faulted metasedimentary and metavolcanic rocks. Thought to be remnants of overlying rocks that remained as survivors between individual plutons, they are especially abundant in the high eastern parts of the two parks.

The present appearance of Sequoia and Kings Canyon National Parks is that of a broad west-sloping plateau deeply cut by branching canyons. In this highest part of the Sierra Nevada the canyons are very deep: Kings Canyon, in places nearly 2500 meters (8000 feet) deep, surpasses Grand Canyon in depth, though it does not rival it in length. The long, straight canyon of the Kern River is more than 1600 meters (about 1 mile) deep. The three main rivers of these parks—the Kings, Kaweah, and Kern Rivers—never reach the sea; their waters sink into or evaporate from the Central Valley to the west. Streams that drain the drier east side of the Sierra, outside park boundaries, flow into the Owens Valley. Their water now is transported by aquaduct to Los Angeles, leaving the Owens Valley a virtual desert.

Uplift of the range seems to have taken place bit by bit. According to many geologists, with each uplift (which may have lasted for millions of years), revitalized rivers cut deep, narrow, usually V-shaped canyons, in time broadened them, and with their tributaries eventually wore the mountains down to rolling hills and shallow valleys. With the next uplift the cycle was repeated, with a new erosion surface almost (but not quite) obliterating the old. In places, particularly in the Kern River drainage, remaining fragments of several ancient erosion surfaces can be recognized. The top of Mount Whitney is thought by adherents of this interpretation to be a remnant of the lowland surface that existed before any uplift took place. One of the between-uplift surfaces appears today as the hilly plateau around Giant Forest, now about 2000 meters (6500 feet) in elevation. Some erosion surface remnants, too windswept to collect deep snow, escaped glaciation in Pleistocene time; others were buried beneath icecaps that blanketed most of the range above about 2700 meters (9000 feet).

New research on the origin of these erosion surfaces suggests that some or all of them may be products of normal erosional processes in granite. The rock, after all, formed deep in the crust, where pressures are almost unimaginable, and is now exposed at the surface, where much lower pressures allow expanding and cracking of the rock, and perhaps breaking apart of individual mineral grains—certainly weakening the rock and making it more susceptible to erosion.

Except in their lower reaches, canyons in this part of the range show plentiful evidence of glacial action: moraines, steep-walled glacial troughs, waterfalls leaping from hanging valleys, rounded roches moutonnées, isolated glacier-dropped boulders (erratics), polished rock surfaces and glacial striae, high tarns and glacial lakes in rockbound basins. Spectacular arrays of matterhorn peaks and narrow arêtes, gouged by cuplike cirques, mark the Whitney Crest and the Great Western Divide.

A prominent feature of these parks, in terms of weathering and erosion, are great granite knobs such as Moro Rock, Beetle Rock, and the Baldys. Created and maintained by weathering processes, they were not, as is often surmised, rounded by glaciers. Rather, curving "leaves" of rock break away in great scales along pressure-release joints.

Many other joints in the granite developed as the original magma solidified. Cooling from the outside in, outer portions of the plutons must have so-

Snow lies late in glacial troughs below the saw-toothed ridge of the Great Western Divide, here seen from Moro Rock.

National Park Service photo

Kern Canyon, accessible only by trail, is a magnificent example of a U-shaped glacial trough. The glaciers that carved it followed a pre-existing stream valley which in turn was eroded along the straight line of a major fault.

lidified while internal portions were still molten. As the molten cores continued to push up and out, the outer, solidified parts of the plutons broke in parallel joint sets. Fluids distilling from the molten cores filled some joints, there to harden into the veins and dikes that now ornament many rock surfaces.

Closely spaced vertical or oblique joints control erosion in many parts of the park, as do faults. The long, nearly straight valley of the Kern River, for instance, follows a major north-south fault zone 130 kilometers (80 miles) long, a zone of shattered rock easily eroded by the river and by glaciers that at times followed the river's course.

GEOLOGIC HISTORY

Paleozoic and Mesozoic Eras. As far as we are able to tell, the history of the Sierra Nevada itself began in Jurassic time, about 200 million years ago, and was occasioned by the parting of North America and Europe. Before then, the Sierra region lay west of the great supercontinent Pangaea, in a sea in which were deposited many layers of sedimentary and volcanic rocks. As the North American Plate drifted westward over the Pacific Plate, these sedimentary and volcanic rocks of the sea floor were bent, broken, and squeezed into the metamorphic rocks of a folded mountain range.

Rocks near the new western edge of the continent, dragged downward with the Pacific Plate, melted at depth. Lighter and less dense than the basalt of the Pacific Plate, bubbles of melted rock pushed upward, enveloping and in part consuming the folded rock of the earlier mountains. Without reaching the surface, each bubble of magma slowly cooled and solidified as one of the plutons of the Sierra Batholith. By the end of Cretaceous time, the great Batholith was in place.

Cenozoic Era. Through most of Tertiary time erosion took over, slowly flattening the mountainous terrain, laying bare the granite plutons and the stringers and sheets of older metamorphic rock between them. By 10 million years ago, the Sierra region had been worn down into gently rolling hills and ridges similar in form (though not in geologic makeup) to today's Appalachians, clothed, in the warm, moist climate of the time, in lush sequoia forests.

Uplift began again about 10 million years ago in the northern Sierra, only 3 to 2 million years ago here in the southern Sierra. As the eastern edge of the great mountain block lifted, tilting its flat surface westward, the climate continued to cool (perhaps in part because mountains *were* rising in many parts of the world). In northern parts of the North American and Eurasian continents, more snow fell in winter than melted in summer, and soon great ice caps began to flow across the land. The Sierra highlands, south of the continental glaciers, spawned smaller ice caps, from which long fingers of flowing ice crept down existing stream valleys.

Both streams and glaciers shaped the landforms we see today. Here as elsewhere, glaciation was a cyclic thing, with glacial advances alternating with warm interglacial stages during which the climate was as warm as it is now, or even warmer. In all, glaciation in the Sierra lasted from about 2.7 million years ago to about 10,000 years ago—having begun at essentially the same time as uplift of the range. During this time, an estimated 14 kilometers (9 miles) of rock were eroded off the top of the rising range, carried into the Central Valley to the west and the Nevada deserts to the east.

The cold climate of glacial times nearly destroyed the magnificent sequoia forests. Today only sheltered pockets of *Sequoiadendron giganteum* remain in the mountains; *Sequoia sempervirens,* the coast redwood, exists now only in the foggy northern California coastal belt. One other species, *Metasequoia glyptostroboides,* the dawn redwood, survives in China and in some western gardens. Man himself came close to destroying these remaining representatives of an ancient family.

The Sierra Nevada continues to rise: Not all California earthquakes can be credited to horizontal movement on the San Andreas Fault! Vertical movement along other faults, especially those east of the range, has occurred in historic time. An 1872 earthquake—the severest California has known—destroyed the town of Lone Pine, just east of the range on the Sierra Nevada Fault; new fault scarps showed that the mountain front had risen 4 meters (13 feet). Numerous earthquakes, as well as movement along previously unrecognized faults, have occurred in the Mammoth Lakes region since 1979.

BEHIND THE SCENES

Beetle Rock and Sunset Rocks. Several of the rounded domes brought about by exfoliation along pressure-release joints occur along the rim of Marble Fork Canyon west of the Giant Forest. Here you can see the leaflike scaling that keeps these rocks rounded. Some scales are less than 1 centimeter (½ inch) thick; elsewhere large slabs a meter (1 yard) or more in thickness have loosened along prominent curving joints that parallel the rock surface.

The granite in this area, light gray, fairly coarse-grained, and containing crystals of feldspar, quartz, and hornblende, is part of the Moro Rock Pluton. It displays few joints other than the pressure-release

Interesting channelways develop along joints where the granite has apparently been weakened by groundwater that seeped through the narrow cracks.

ones. White or light gray pegmatite dikes that cross the rock have about the same mineral composition, but their crystals are larger. Follow one of these veins—you'll probably find that it soon branches or pinches out. Interesting flat-bottomed channelways run along some of the joints, to end in tiny stairstepped hollows, sites of miniwaterfalls during rains. Water standing in rockbound hollows becomes slightly acid with the metabolic products of tiny pond-dwelling plants and animals; the acid acts to dissolve rock minerals, loosening the grains and thereby helping to enlarge the little ponds.

There are inclusions of diorite in the granite also. Some of them, harder than the surrounding granite, project as small knobs.

Lichens, pioneers of the plant world, coat many rock surfaces. Secreting tiny amounts of acid, these fungi-algae partnerships help in breaking down the rock to which they are attached, and hence figure in early phases of soil development. Mosses grow in crevices and hollows, furthering the production of soil. Even trees contribute: Sprouting from seeds caught in moisture-holding crevices, they force rocks apart with their growing roots.

The western tip of Sunset Rock overlooks the Marble Fork of the Kaweah River, where Marble Falls plunge over a marble ledge. The blue-gray marble, once limestone on the floor of the sea, is part of the north-south band of metamorphic rocks that extends across this part of Sequoia National Park. To see it up close, go to Crystal Cave.

The ridge west of Beetle and Sunset Rocks is made of the metamorphic rocks of a roof pendant. Little Baldy, visible to the north, is another granite dome.

Big and Little Baldy. Both accessible by trail, these granite knobs are rounded along curving pressure-release joints. They are among the many such domes here and elsewhere in the Sierra Nevada. On the west side of Little Baldy, rockslides along steep, straight pressure-release joints have created unusual rock-avalanche chutes.

Crescent Meadow. Bare knobs of granite rise along the trail to this mountain meadow. Like Moro Rock, they are part of the Moro Rock Pluton. Notice the coarse sand into which the granite weathers—sand made up, like the granite, of both quartz and feldspar grains. Mica and hornblende have altered to clay minerals and hematite, the latter giving the sand its rusty color. The meadow itself, now filled with fallen trees and marsh vegetation, was once a small lake formed in a shallow depression on the poorly draining west slope of the Sierra block.

Crystal Cave. Small but attractive, this cavern is etched in marble produced by recrystallization of limestone. Underground water, carrying weak acids derived from atmospheric carbon dioxide and soil, seeped through joints and along sedimentary layers, dissolving as it went minute amounts of calcite, the principal mineral of limestone and marble. Below the water table, where acidified water completely filled all openings in the rock, long-continued solution created sizeable underground chambers. Later, as the climate changed or as streams cut down, seeping water charged with calcium carbonate dripped and trickled through empty cavern chambers. As droplets hanging from the ceilings lost some of their carbon dioxide, tiny rings of calcium carbonate were deposited. As dripping continued the rings grew into slender tubules and long stalactites—a process that is still going on. Water that drips to the cavern floor builds, with its splashing, massive stalagmites. Where stalacite and stalagmite meet, columns form. Water sheeting down chamber walls leaves a variety of flowstone features—veils, curtains, and "waterfalls." Parts of the cave ceiling show the original limestone layering, marked with joints decorated with rows of small stalactites.

Crescent Meadow, once a small lake, will eventually support forest trees.

Giant sequoias, once far more widespread, are now confined to sheltered Sierra habitats with just the right temperature and rainfall.

Informative signs along the trail make good resting places on the way back from the cave. Cascade Creek tumbles over vertically tilted schist and gneiss, metamorphic rocks that with the marble are parts of the elongate roof pendant that crosses the western part of the park.

Generals Highway: Ash Mountain to Grant Grove. Entering the park near the Ash Mountain park headquarters, the highway climbs at first through dark gray gabbro and diorite of one of the Sierra plutons. Near Hospital Rock Picnic Area it comes into a band of metamorphic rocks—one of the Sierra roof pendants—that stretches north across the western part of the park. With the change in rock type, soils change from light orange-brown sand to darker brown silt. Both types of soil are tinged to different degrees with red hematite, product of oxidation of iron-bearing minerals. The band of metamorphic rocks is particularly well exposed along the Middle Fork near Tunnel Rock. There the intricate, taffylike folds that characterize these rocks are beautifully displayed.

Among the metamorphic rocks, a broad band of blue-gray and white marble is easily recognized: It forms Paradise Ridge south of the Kaweah River and extends northward through the picnic area, across Amphitheatre Point, and to Marble Falls and beyond. Standing vertically, the marble is sandwiched between layers of quartzite, slate, and shiny mica schist. Because these metamorphic rocks are sliced by many joints, they are much easier to excavate than is the massive granite nearby. The highway was therefore routed in short switchbacks up the metamorphic ridge. At the top of the switchbacks the highway reenters the granite.

From vantage points along the road you'll see the towering dome of Moro Rock. Look carefully; there are people on top. Both Moro Rock and Hospital Rock are granite domes created by scaling away along pressure-release joints. Most of the rubble below has fallen from Moro Rock.

In the Giant Forest you'll probably want to spend most of the time looking at trees. However there are some geologic features here, too. The giant sequoias all seem to grow on mounds—small hills produced as enlarging roots push up the soil. The steep mounds tend to slough off plant debris—foliage, branches, and bark—and thereby to some degree protect the great trees from fire. Boggy ponds, features caused by irregularities on the gently sloping, poorly drained Sierra west slope, eventually fill in to become meadows. The meadows in turn are gradually taken over by the forest.

Between Giant Forest and the Lodgepole Visitor Center the road cuts through bouldery gravels of old moraines. Soil developing from this material has smoothed the slopes. Curved tree trunks show that soil, creeping slowly down the slopes, tilted them when they were young. The trees straightened as they grew larger and as their roots became deeper and stronger. Snow creep as well as soil creep can tilt pliable young trees.

Near Lodgepole several low granite domes project through the soil mantle. On the domes and in roadcuts you'll see some of the curving pressure-release joints that dictate the shape of these domes.

North of Lodgepole, beyond the bridge over the Marble Fork of the Kaweah River, the granite along the highway comes in many shades of gray, depending on the proportion of the dark minerals hornblende and biotite. Some virtually lacks dark minerals—it is almost pure quartz and feldspar. In places the granite contains large angular to rounded inclusions of dark gray diorite.

This part of the highway route is close to the west-tilted surface of the Sierra block, an unglaciated surface that has seen long, deep weathering, with less removal of the products of its weathering than in the higher, glaciated parts of the mountains. As biotite and other unstable minerals weather, bonds between quartz and feldspar crystals weaken, and the rock turns gradually into sandy "rotten granite" or grus, so soft that it can be dug with a shovel. Since deep weathering takes place along underground joints as well as along those at the surface, angular blocks of granite become rounded and surrounded with grus below as well as at the surface.

Redwood Mountain, the long ridge extending south of the Generals Highway opposite Quail Flat, is the northern extension of the roof pendant of metamorphic rock that you saw at Hospital Rock Picnic Area and Amphitheatre Point. The largest

When jointed granite weathers underground, mica minerals decompose, loosening other mineral grains. Granite blocks gradually become rounder in a process known as spheroidal weathering. A roadcut exposes this example.

National Park Service photo

Seen from the High Sierra Trail, the Great Western Divide is a rugged, glacier-shaped rock wilderness, a land of cirques, arêtes, and steep U-shaped valleys.

still extant grove of giant sequoias is on Redwood Mountain and in Redwood Canyon below it.

By the time the highway reaches the Grant Grove it is back in the granite.

High Sierra Trail. One of the best routes to the "backcountry," this trail runs along the south rim of the canyon of the Middle Fork of the Kaweah, with stunning views, for even the casual walker, over the canyon to the high peaks of the Great Western Divide. Ultimately it climbs the beautiful upper valley of the Middle Fork, a "textbook" glacial trough known as Valhalla, complete with rockbound glacial lakes and many examples of glacial erosion. Watch along the route for other glacial features, which are better preserved here in the high country because receding glaciers remained here longest.

Crossing the Great Western Divide at Kaweah Gap, the trail drops into the equally "textbook" valley of Big Arroyo, a tributary of the Kern River. Peaks east of Kaweah Gap, at the north end of Kaweah Peaks Ridge and farther north to Triple Divide Peak, consist of metamorphic rocks of one of the Sierra roof pendants. Their dark tones contrast with the pale gray of granite peaks on either side. Triple Divide Peak marks the point at which drainage basins of the Kaweah, Kings, and Kern Rivers meet.

From Kaweah Gap the trail drops into the valley of Big Arroyo, a tributary of the Kern River. It ascends again to the Chagoopa Plateau, one of the erosion surfaces thought to have been created between Sierra uplift episodes. Other erosion surface remnants can be seen to the east, across the valley of the Kern River. The trail finally descends the steep wall of Kern Canyon, described below.

Kern Canyon. Accessible only by trail, this long, straight canyon follows a major fault zone. Initiated by stream erosion of shattered and disrupted rocks along the fault, the canyon was occupied in Pleistocene time by a valley glacier. Marks of the glacier's passage are in the U-shaped canyon itself, in hanging valleys that enter it, in cirques, polished rock, and striae of the upper basin (most recently occupied by glaciers), and in rounded, ice-quarried rock knobs.

Being within the rainshadow of the Great Western Divide, the Kern River glaciers were not as well nourished as those west of the divide. They clearly did not completely fill the valley. Avalanche chutes scoured out in glacial times terminate abruptly part of the way down the canyon walls, where sliding snow and rock debris reached the ice surface.

Kings Canyon. The drive from Grant Grove to Kings Canyon takes you from the plateaulike west-dipping slope of the Sierra Nevada to the bottom of the deepest canyon in the range. The rocks through which you pass are typical Sierra rocks: assorted granites of the Sierra Nevada Batholith interspersed with roof-pendant metamorphic rocks: quartzite, phyllite, marble, schist, and gneiss, now tilted on end. Both granite and metamorphic rocks are cut by many joints.

The highway enters the park just below Cedar Grove, having come upriver through the spectacular V-shaped gorge of the South Fork of the Kings River, in places nearly 2500 meters (8000 feet) deep. Unfortunately, such is the shape of the canyon walls that from the bottom you are not really aware of its depth, as you are in Yosemite Valley.

Above Cedar Grove the canyon profile changes from the V shape of the river-cut gorge to the

National Park Service photo

The Kern River flows through a magnificent trough carved by Ice Age glaciers.

National Park Service photo

West of Cedar Grove, Kings Canyon is distinctly V-shaped. Trough-cutting glaciers did not reach this area.

troughlike U shape of a glaciated valley. A boulder-strewn terminal moraine near Cedar Grove marks the farthest glacial advance. On either side of the valley floor, especially where joints are few, the granite walls approach the vertical, and massive granite domes mark the skyline. In contrast, where joints are closely spaced the cliffs are craggy and less steep.

An oxbow lake—isolated remnant of a Kings River meander—can be seen from the Cedar Grove Motor Trail, as can many other geologic features, some described in the trail leaflet. Farther up the canyon are pleasant meadows and forested flats, some of them on the site of an infilled lake dammed by the terminal moraine at Cedar Grove. Sentinel and North Domes, both products of scaling along pressure-release joints, tower above the glaciated trough. In spring and early summer little waterfalls plunge from hanging valleys.

Mist Falls. The trail to Mist Falls begins at the end of the Kings Canyon road. With good views of the glacier-carved cliffs that tower above the confluence of Bubbs Creek and the South Fork, it curves north around Buck Peak up the South Fork. Granite close to the trail contains large, rectangular crystals of feldspar, some 2 centimeters (1 inch) or more in length. Many dikes cross the granite, some distinctly banded with crystals of different sizes. Where the trail crosses the bare granite, watch for glacial striae and polish.

The falls themselves burst from a narrow passage through massive, relatively unjointed granite that seems to have been able to withstand the might of the glaciers. What a crush of ice there must have been here in glacial times, with ice blocks tumbling where the water tumbles now. The trail continues up into Paradise Valley on the South Fork, and then up Woods Creek to the John Muir Trail just below the crest of the range.

Moro Rock. The climb up Moro Rock may leave you breathless, but it is well worth while. The rock itself is made up of massive, relatively unjointed Moro Rock Granite. Coarse, chunky crystals of feldspar and quartz are liberally peppered with rods of hornblende and tiny flakes of muscovite (white mica). Here and there the granite is marked with inclusions of darker diorite. The rock is thickly coated with lichens, and its moist crevices and cracks harbor mosses, flowering plants, and a few stunted trees.

You need not go all the way to the top of the rock to see these features or to observe the cabbagelike layers, separated by pressure-release joints, responsible for the shape of the rock. Curving rock slabs break away and add to the talus below. Similar talus, some of it at the bases of long rockslide-avalanche chutes, foots the slopes across the valley.

From landings along the trail, or from the summit, look east into the upper basin of the Middle Fork of the Kaweah River, backed by the sawtooth summits of the Great Western Divide, with its cuplike cirques and sharp arêtes. Note the difference between the steep, U-shaped glacial trough upstream and the V-shaped river canyon 1200 meters (4000 feet) below you. The greatest of the glaciers to occupy this valley terminated almost directly below Moro Rock.

The Kaweah River flowed originally into Tule Lake in California's Central Valley. Now most of its flow is diverted into irrigation canals and aquaducts.

Northwest of Moro Rock the Generals Highway zigzags up a band of brown, densely jointed metamorphic rocks, one of the Sierra roof pendants. The band extends from Paradise Ridge, south of the Middle Fork, northward to Redwood Mountain near Grant Grove. Standing out because of its blue-gray color, a thick bed of limestone, altered to marble, extends northward across the Marble Fork of the Kaweah River: It is the host rock of Crystal Cave.

Mount Whitney. Most of the Mount Whitney climb is outside Sequoia National Park, the summit ridge marking the boundary between the park and national forest land to the east. It provides a good opportunity, however, to look at the imposing east face of the Sierra Nevada. Here the canyons are shorter, steeper, and drier than on the western slope. All streams flow into the Owens Valley and never reach the sea. Coarse granite along the trail is crisscrossed by white veins of quartz and feldspar. Small glacial lakes lie in stairstepped canyons backed by the sheer east face of the Mount Whitney-Mount Muir massif. Parallel avalanche chutes that flute the cliffs do not reach the floor of the cirque, as you might expect, but terminate well above, at a level that marks the former top of glacial ice.

From the Trail Crest divide you can look down, westward, into another deep cirque, occupied now by Hitchcock Lakes. The glaciated landscape spread out before you is the upper end of the Kern River Basin. Across it, 20 kilometers (12 miles) away, rises the Great Western Divide, almost as high as the crest on which you stand. Instead of flowing east or west like other Sierra rivers, the Kern River flows directly south through a long, straight canyon that follows a major fault zone, dividing the High Sierra into two crests.

The smooth, sloping plain west of the final stretch of the trail is thought to be a remnant of the

Moro Rock's massive dome is rounded by exfoliation along pressure-release joints. Figures on summit give scale.

National Park Service photo

ancient lowland erosion surface that existed here before the Sierra fault block began to rise. Long-preserved because not enough snow accumulates on its windswept surface to form glaciers or to melt into streams that might gully it, it has been cut away on the east by headward-eroding streams and glaciers. Its veneer of rock fragments shattered and set at angles by freeze-and-thaw processes protects it from winter gales that long ago blew away all the finer fragments.

Tokopah Falls and Tokopah Valley. Glaciers that flowed down the canyon of the Marble Fork reached well downstream from Lodgepole, to elevations of about 1800 meters (5900 feet). There are three recessional moraines between the main highway and the upper end of the campground, all much disrupted by man. Large moraine boulders, however, border the river and dot the campground. Granite knolls above the bridge, rounded on their upstream sides and plucked away on their downstream sides, are studded with erratics, large boulders of unmatched granite lying just where the receding ice left them.

The bare granite is interesting in itself, revealing north-south joints, fine-grained diorite dikes, and a pegmatite dike with large quartz and feldspar crystals.

Upstream, where glacial ice was thicker, the walls of the valley steepen. In and after wet weather, slender waterfalls plunge from the cliffs of the Watchtower, sheeting down the glacier-scoured rock faces of its lower wall.

Pink orthoclase feldspar colors the granite near Tokopah Falls. Most of the huge angular boulders below the falls were brought by the glaciers from upper Tokopah Valley; some doubtless fell from adjacent cliffs as the supporting ice melted back.

OTHER READING

Oberhansley, F.R. 1946. *Crystal Cave in Sequoia National Park, California.* Sequoia Natural History Assoc., vol. 1, no. 1.

Ross, D.C. 1958. *Igneous and Metamorphic Rocks of parts of Sequoia and Kings Canyon National Parks, California.* California Division of Mines Special Report 53.

Sequoia Natural History Association. 1968. *Moro Rock, Its History, Biology, and Geology.* (Out of print, but available in some libraries.)

Yosemite National Park

Established: 1890
Size: 3079 square kilometers (1189 square miles)
Elevation: 648 to 3979 meters (2127 to 13,053 feet); Yosemite Valley Visitor Center 1219 meters (4000 feet)
Address: P.O. Box 577, Yosemite National Park, California 95389

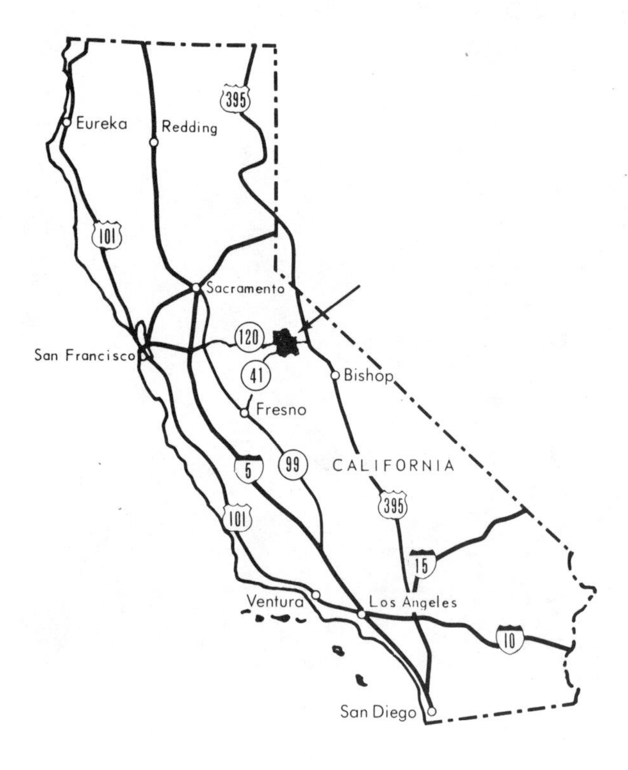

STAR FEATURES

• A glacier-carved valley of exceptional grandeur, its vertical granite walls rising as much as 1200 meters (4000 feet) above the meadow-and-forest valley floor.

• Foaming waterfalls plunging from hanging valleys. Snow-fed, the falls are at their best in late spring and early summer.

• Coarse-grained intrusive rocks of the Sierra Nevada Batholith, with scattered outcrops of older metamorphic rocks that occur between intruding granite plutons of different ages.

• High Sierra country, vast, wild, much of it above 3000 meters (10,000 feet) in elevation, most of it accessible only by trail. The ice-scoured uplands are surmounted by frost-sharpened peaks; glacial features abound.

• Visitor centers, museum, guided hikes, daytime and evening programs. Roadside and trailside displays explain geologic features; available in visitor centers is a roadguide keyed to numbered roadside markers. Topographic maps of Yosemite Valley and the park as a whole are also available at visitor centers, as is a geologic map of Yosemite Valley.

See cover and color pages for additional photographs.

SETTING THE STAGE

Stretching more than half the length of California, the Sierra Nevada is one of the largest fault block ranges in the world. Intensely faulted along

In Yosemite's high country, glacial features abound. Small sharp peaks like Unicorn and Cathedral Peaks, center, jutted above the great icecaps of Pleistocene time. Foreground boulders, glacial erratics, lie atop Pothole Dome, a well rounded roche moutonnée.

The ponderous cliff of El Capitan (left) and snow-fed Bridalveil Falls (right) owe their existence to glaciers of Pleistocene time. In the distance is Half Dome, thought to have projected above the glaciers.

On Fairview Dome, glacial polish shines in the morning sun.

the east side by relatively young faults, the Sierra block as a whole tilts westward, with its sharp eastern edge raised like a giant trapdoor. Its western slope slants gradually into California's Central Valley.

The high, continuous barrier of the Sierra intercepts prevailing westerly winds, capturing their moisture in the form of deep-drifted winter snows. Into the tilted western slope, snowmelt and ice have, through time, carved a legion of chasms—deep, imposing, sheer-walled troughs. Of these, Yosemite, the valley of the Merced River, reigns supreme.

Only about 11 kilometers (7 miles) long and hardly 1 kilometer (0.6 mile) wide (a width about equal to its depth), Yosemite Valley is one of the loveliest spots on earth. Its meadowed floor and handsome granite cliffs, its majestic waterfalls and foaming cascades, its mirrorlike pools and great rockfalls tell an eloquent story of uplift and erosion, of changing climates, of winters and summers and floods of spring, a story verified by features in the uplands that surround the valley itself.

The floors of Yosemite Valley and its near neighbors contain glacial moraines, and behind them flat-lying lake deposits as much as 700 meters (2300 feet) deep. Glacial deposits also dot the backcountry, the High Sierra, where barren ice-polished rock alternates with rubbly moraines and scattered glacial lakes. Glaciers were major sculptors in all of Yosemite: Glacial ice formerly covered close to 70 percent of the present national park. Everywhere its hand is evident: in spectacular cliffs, hanging valleys, high-plunging waterfalls, and stairstepped canyon floors.

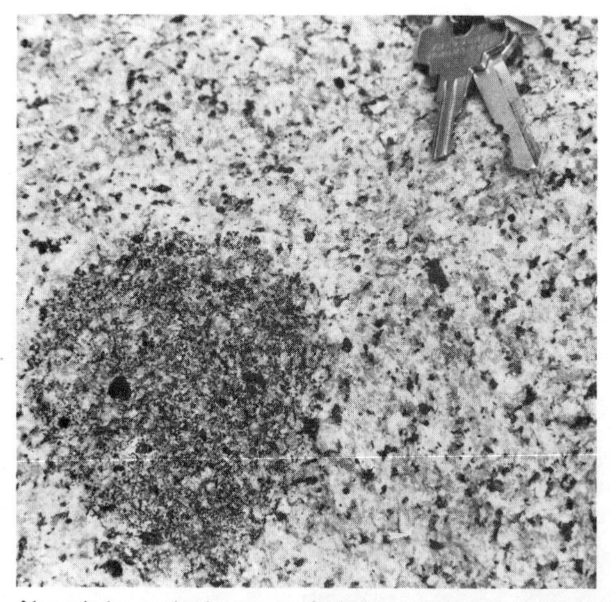

Yosemite's granite is composed of glassy quartz (gray in this photo), white feldspar, black hornblende, and mica. Finer, darker inclusions like the one at the left are common in some of the plutons.

Practically all the rocks in Yosemite are intrusive igneous rocks of the Sierra Nevada Batholith, granite and its kin. True granite is composed primarily of quartz and feldspar, with sprinkles of dark minerals such as biotite and hornblende. Some of the granite-like rocks can be classed as granodiorite or as quartz monzonite, depending on the proportion of different minerals of the feldspar group. However, since these three rock types are all light gray in color and look very much alike, here as in much geologic literature they are all referred to as granite.

Other intrusive rocks seen here are diorite, much darker gray because of more abundant biotite and hornblende and the absence or near absence of quartz. Gray veins of diorite or white ones of quartz and feldspar mark many rock faces, particularly in the Hetch Hetchy and Tuolumne Meadows areas.

In the Sierra Nevada as a whole more than 100 intrusions, individually called plutons, have been identified and described. They are all lumped under the general term Sierra Nevada Batholith. Each of the plutons is thought to represent a different igneous intrusion, a different balloon of magma that pushed its way upward as a result of subduction and remelting at the edge of the continent.

Some of the plutons seem stronger, more massive, than others, largely because they have fewer joints brought about by stresses and strains of uplift and mountain-building. Both El Capitan and Half Dome Granites, for instance, are fairly free of joints, and so stand up as the highest, sheerest cliffs bordering Yosemite Valley: El Capitan's near-vertical face is 1000 meters (3000 feet) high.

As you will note, though, the massive El Capitan and Half Dome Granites do display some vertical joints—joints that helped the glaciers to whittle the valley walls. Half Dome's northeast shoulder displays closely clustered vertical joints along which its sheer north face was cut away. Other joints have also been important in shaping the rock masses—those that parallel exposed rock faces, whether sheer or curved. Intrusive igneous rocks, we must remember, cooled and hardened far below the surface, under many thousands of meters of overlying rock, with almost unimaginable pressures from the weight of rocks above them. Long after they cooled, the Earth's crust was pushed and lifted, and overlying rocks were eroded away. As pressures lessened, the igneous rocks expanded—ever so slightly, but enough to open fractures parallel to their surfaces. On the many rock domes of Yosemite, weathering along these curving pressure-release joints has gradually loosened great curving "leaves" of rock, peeling

Ray Strauss photo

The east side of Half Dome shows well the pressure-release joints that govern its shape.

Pressure-release joints show up also along the Tuolumne River below Tuolumne Meadows. They may in part be due to release from weight of the heavy icecap that once covered this area.

them away like onion layers, a process that is still going on.

Other granitic rocks of Yosemite, sliced by closely spaced vertical, diagonal, and horizontal joints, weather into jagged pinnacles rather than rounded domes. As moisture in joints freezes, thaws, and refreezes again and again, these rocks are broken and pried apart. Loosened from the parent mass, many fragments fall to the valley floor, accumulating there in rough rock piles or talus. Erosion along vertical and slanting joint sets may produce rows of sibling peaks like the Three Brothers.

In the eastern part of the park, along the Sierra crest, are volcanic rocks metamorphosed to greenstone, as well as some strongly altered sedimentary rocks, some of which contain distorted but recognizable Paleozoic fossils.

Interesting features in the Sierra are the so-called roof pendants, irregular pockets of metamorphic rock thought to have hung down into and between the plutons of the Sierra Nevada Batholith. The altered sedimentary rocks are predominantly Paleozoic; the altered volcanic rocks are predominantly Mesozoic. All these rocks are strongly and complexly folded and faulted. Careful study of several roof pendants shows that Mesozoic folds and faults are superimposed on late Paleozoic folds and faults, and these in turn on mid-Paleozoic folds and faults, making of the roof pendants complex three-dimensional jigsaw puzzles. With plenty of geologic detective work, though, the roof pendants, together with less altered sedimentary and volcanic rocks of the western foothills, have revealed some of the early geologic history of this area.

GEOLOGIC HISTORY

Paleozoic Era. Through most of Paleozoic time the Yosemite region lay along the western edge of the supercontinent of Pangaea. As the wide, flat coastal plain warped and sank below sea level and rose again, thick layers of sedimentary and volcanic rock were spread out upon it—some marine, some deposited above the sea as floodplains and deltas. There seem to have been some complicated faulting, folding, and mountain-building late in Paleozoic time, as well as long periods of erosion.

Mesozoic Era. Sediments and great volumes of volcanic rocks continued to be deposited in Triassic and early Jurassic time. Then, between 210 and 80 million years ago, westward drift of the newly formed North American continent brought changes to its western margin. Colliding with and overriding the East Pacific Plate, the North American Plate bent and crumpled, and saw upwelling of the plutons that now form the core of the present Sierra Nevada. Their light color and mineral composition tell us that the huge bubbles of molten rock were of continental material melted by magmas from the East Pacific Plate as it plunged beneath the continent.

The plutons apparently developed along three belts, in three long-lasting pulses, progressing from east to west. Plutons of the first pulse—mostly small, dark ones composed of diorite—lie farthest east, along the Sierra crest and in some of the smaller ranges of Nevada. Those of the second pulse now form the central Sierra. Yosemite Valley lies in rocks of the third pulse.

As they forced or melted their upward way, the plutons twisted and bent overlying sedimentary and volcanic rocks of Paleozoic and Mesozoic age, heating and altering them. Isolated fragments of older rock remained between the plutons as roof pendants, as we have seen. The later plutons also loosened and incorporated small fragments of the older diorite, which now appear as dark gray inclusions embedded in lighter granite.

The new plutons cooled gradually, well below the surface, under conditions of very high temperature and pressure. (The oldest, easternmost plutons may have been the shallowest; they are associated with volcanic rocks of similar composition, which may represent the same magma broken through to the surface.) Shrinking as they cooled, the intrusive rocks cracked, and the cracks filled with fluids seeping from the main crystallizing mass, creating the many veins that can be seen cutting through the rocks today.

A long period of erosion followed, lasting well into the Cenozoic Era.

Cenozoic Era. By 55 million years ago, most of the mountains formed in Mesozoic time were gone, leveled by erosion. The plutons had been laid bare. Around 20 million years ago the Earth's crust began to crack and break along the west edge of the continent. Some of the resulting blocks, formed at different times during the preceding 20 million years, were tilted or up-ended to form separate mountain ranges; others sank to form basins between the ranges. The largest of the lifted units—a block more than 600 kilometers (400 miles) long and up to 120 kilometers (80 miles) wide—today forms the Sierra Nevada. Its eastern margin was lifted high along the Sierra Nevada Fault; its western margin hides deep under the San Joaquin—Sacramento Valley.

There is ample evidence that the rise of the great block occurred slowly and in many increments, and is still going on today. As the range rose, west-flowing streams cut deep valleys into it, then broadened the valleys. After another uplift, streams

Dark inclusions of diorite spot this great granite boulder moved downstream by Yosemite's largest glacier.

A dark gray diorite vein separates two kinds of granite in this close-up photograph.

A forked vein cuts through fine-grained gray granite in this roadcut exposure. A few dark gray inclusions also mark the granite. Four vertical scars are drill holes from blasting.

cut downward again, creating new valleys within the old.

Pleistocene time brought other changes. The climate became wetter and colder. More snow fell during long, cold winters than could melt in short, cool summers, and gradually the northern part of the Northern Hemisphere was overcome with a great blanket of ice.

In the Sierra, its high country catching moisture from Pacific winds, a large ice cap covered the uplands. Like fringes of the icy plateau, valley glaciers flowed down existing stream valleys. Loaded with rock fragments plucked from the high peaks, the ice ground away spurs and scoured, deepened, and straightened the valleys, carving the Yosemite we know today.

At least three times the glaciers advanced and retreated. Glacial advances in Yosemite and in the Sierra as a whole have been called by a variety of names. For our purposes, let's refer to the three main advances by the positions of their terminal moraines, and call them the El Portal, Bridalveil, and Nevada Fall Glaciers. Probably there were earlier stages as well, but the El Portal Glacier effectively erased evidence of them.

The El Portal Glacier, which reached its maximum size about 75,000–60,000 years ago, was by far the greatest and mightiest river of ice, in places 1200 meters (4000 feet) thick. As it ground through Yosemite Valley, filling it from rim to rim, Glacier Point and North Dome felt ice across their summits, though Half Dome and El Capitan projected above it. Formed by the union of glaciers from Tenaya Canyon and the upper valley of the Merced River, the great glacier also received small tributary flows from both sides. Since with its enormous mass it cut downward far faster than did its little tributary glaciers, it brought about the difference in elevation between Yosemite Valley and its tributary canyons—the difference responsible for today's magnificent waterfalls. Flowing west down the canyon of the Merced River, the glacier reached a point a few kilometers downstream from El Portal.

Even at its maximum extent 45,000 years ago, the Bridalveil Glacier was no match for its predecessor. Its terminal moraine remains as rubbly rockpiles between Bridalveil Fall and El Capitan. As it flowed through Yosemite Valley, never reaching the tops of the cliffs, it tidied up the scattered boulders and recessional moraines left by its predecessor, and smoothed and sharpened the cliffs on either side of the valley. Most significantly, its terminal moraine, near Bridalveil Fall, dammed the valley. When the ice retreated, the valley was the site of a new lake, Lake Yosemite.

Ray Strauss photo

Throughout Yosemite one comes across examples of glacial polish and the striae that result as rock scrapes across rock.

The Nevada Fall Glacier reached its maximum extent 20,000 years ago. Though it never reached Yosemite Valley's floor, it did put finishing touches on the upper valley of the Merced River, known today as Little Yosemite, helping to shape the stepped giant staircase of that valley. And with a similar tongue of ice that flowed down Tenaya Canyon, it contributed to the volume of coarse and fine rock debris that in time filled in Lake Yosemite to form the almost level floor of the valley as we see it today.

One small glacier exists in the park today, a double-lobed one on the flanks of Mount Lyell. Several others lie east of the divide, just outside the park. These small masses of ice, glaciers in that they flow and grind up rock and release its chalky substance to streams, came into existence only 4000–3500 years ago, during a cold-climate cycle known as the Little Ice Age. The tiny glaciers are shrinking now, as are those elsewhere in the Sierra and in many other parts of the world.

BEHIND THE SCENES

Bridalveil Fall. Regularly flowing all summer long, this beautiful fall leaps 189 meters (620 feet) from a high, classic hanging valley shaped by a tributary of the El Portal Glacier. Studies of hanging valleys such as this one have helped geologists outline the history of the Yosemite region. By projecting the line of the tributary valley floors outward over the main valley, they can estimate the size and shape of the preglacial Merced River canyon.

The rock behind Bridalveil Fall represents two of Yosemite's seven plutons: the upper, smooth-surfaced Bridalveil Granite, and the lower, rougher-surfaced Leaning Tower Granite.

The cliff over which the fall plunges is largely a product of the El Portal Glacier, the only glacier to reach farther downstream. Some of the boulders in the terminal moraine of the Bridalveil Glacier—distinctive granite boulders with large white feldspar crystals—were carried here from the Tenaya Lake area 30 kilometers (18 miles) upstream.

Since the cliff was carved, it has been gradually undermined by the waterfall that plunges over it. Constantly wet with spray, minerals like mica decompose. Their decomposition loosens other mineral grains, which then wash away. Winter frost wedges rocks apart along joints, and the force of falling and splashing water further pries rock away from the cliff.

The Bridalveil Glacier reached a point just downstream from the fall, at the rocky, hummocky terminal moraine that closes the downstream end of Bridalveil Meadow. What we see is actually only the top of the moraine, the upper edge of a 600-meter-high (2000-foot) natural dam that for a time backed up Lake Yosemite.

Across the valley from Bridalveil Fall, dark gray diorite, older and more densely jointed than its neighbor the El Capitan Granite, forms rockslides of angular, broken rock.

El Capitan (clifftop elevation 2146 meters, 7042 feet). El Capitan forms, with Cathedral Rocks, the gateway to Yosemite Valley. The mighty cliff rises 883 meters (nearly 3000 feet) above the valley floor. It developed where the El Capitan Granite, massive and relatively unjointed, cuts across the valley to Cathedral Rocks. Both El Capitan and Cathedral Rocks owe their present shape to the El Portal Glacier; their cliffs are maintained now by exfoliation along pressure-release joints that approximately parallel the rock faces.

Within the massive granite are dark veins of gray diorite—not to be confused with vertical streaks of lichens established where water seeps to the surface. The so-called "map of North America" on El Capitan's eastern face is a larger diorite mass. Since it cuts across the granite, the diorite is the younger. A few thin white veins are still younger; they cut across the diorite. Just west of El Capitan, a large diorite pluton, densely jointed and therefore a poor cliff-former, breaks up into the Rockslides. This diorite is older than the El Capitan Granite.

None of the Yosemite Valley glaciers flowed over the summit of El Capitan, which is about 150 meters (500 feet) higher than the top of the great cliff. The El Portal Glacier, however, is thought to have reached just to the top of the cliff.

Glacier Point (elevation 2199 meters, 7214 feet). From Glacier Point the upper part of Yosemite Valley is laid out before your eyes—one of the most magnificent views you will ever see. To the north, and practically straight down, the Merced River winds across Yosemite Valley's nearly level floor. To the southeast, Grizzly Peak, Liberty Cap, and the north slopes of Mount Starr King frame Little Yosemite Valley. To the northeast, Half Dome towers over the bare-rock walls of Tenaya Canyon, and North Dome rises above Royal Arches and the Washington Column.

El Portal and Bridalveil Glaciers, creeping down Tenaya Canyon and Little Yosemite Valley, merged below Glacier Point.

Distant uplands seen from here are the site of the great ice caps of Pleistocene time. With the rapid uplift that immediately preceded glaciation, Tenaya Creek and the Merced River deepened their canyons and created a new, more rugged topography

Royal Arches were created as masses of undermined, unsupported granite fell away from the north valley wall. Under the stress of gravity, slabs of massive, unjointed rock break away in long curves.

that, as the climate grew colder, guided the flowing tongues of glacial ice born of the High Sierra ice cap.

Nearly 1000 meters (3200 feet) above the valley floor, Glacier Point was overridden by ice of the El Portal Glacier. Evidence lies in glacial striae and polish here on the point itself, as well as in glacial erratics—boulders of distinctly different rock types—near its summit and on the summit of nearby Sentinel Dome.

Half Dome (elevation 2695 meters, 8842 feet). Composed of massive Half Dome Granite, this great rock monument is a product of both glaciation and pressure-release jointing. The imposing precipice of its north face was established by glacial plucking along vertical joints, some of them visible on the low shoulder about two-thirds of the way up this face. Although early authors supposed that Half Dome was first rounded by glacial ice and then sliced in two by another glacier, recent research has shown that Half Dome's summit rose 150 meters (500 feet) above the El Portal Glacier, and much more of course above later, smaller glaciers. Its rounded south side is a product of scaling, or exfoliation, along pressure-release joints.

Half Dome's rounded summit was produced by spalling along pressure-release joints. Its north face steepened as glaciers chiseled an intensely jointed part of the rock. Closely spaced vertical joints can be seen in the lower part of this photograph.

The Half Dome Pluton, particularly prone to erode into domes, extends far north, east, and south of Half Dome itself, contributing its massive strength to North Dome, Basket Dome, Grizzly Peak, Liberty Cap, and the steep, bare-rock walls of Tenaya Canyon.

Hetch Hetchy. Now lake-filled, Hetch Hetchy Valley, the valley of the Tuolumne River, is similar in origin to Yosemite Valley. Walled by glacier-carved cliffs, it is the scene of a number of lovely waterfalls that plunge from hanging valleys left by tributary glaciers. Tueeulala Falls plunge about 300 meters (1000 feet). Smooth granite surfaces of El Capitan Granite show the decorative patterns of many veins.

Upstream is the Grand Canyon of the Tuolumne River, a long, U-shaped glacial trough heading in Tuolumne Meadows and the high country near Tioga Pass.

Little Yosemite Valley. Little Yosemite Valley displays a classic example of a glacial stairway, where flowing ice scoured out nearly horizontal stairway "treads" and nearly vertical "risers." The positions of treads and risers are governed by spacing of joints in the granite: Where joints are closely spaced the glaciers plucked away the rock quite easily, to form the vertical risers. Where joints are sparse, plucking was less effective, so broad treads developed. Joints also affect the width of the glacial valley, which is narrowest where the rock is massive, widest where joints are close together.

Mariposa Grove. Sequoias are "living fossils," descendants of an ancient lineage. The oldest known sequoias, fossils from rocks of Cretaceous age, were contemporaries of the dinosaurs. Only three species survive today: *Sequoiadendron giganteum,* the giant sequoia, which occurs in Yosemite, Sequoia, and Kings Canyon National Parks as well as elsewhere in the Sierra Nevada, *Sequoia sempervirens* or the coast redwood, of Redwood National Park and other parts of California's Coast Range, and *Metasequoia glyptostroboides,* the dawn redwood, a native of China that has now found its way into American gardens. Yosemite's giant sequoias range to at least 3000 years old.

Merced River Canyon. Some of the oldest rocks in the Yosemite region occur outside the national park, in Merced Canyon downstream from El Portal. Tightly folded Paleozoic metasedimentary rocks—schist, quartzite, and marble—are well exposed along the stream and highway. Remnants of the lowest moraine of the El Portal Glacier lie a short distance below El Portal, also outside the park. Patches of coarse gravel with well rounded cobbles and boulders are stream-carried glacial outwash deposits.

East of El Portal the deep, narrow, V-shaped canyon of the Merced River is walled with light gray granite patterned with veins and irregular inclusions of diorite. The present stream canyon has been cut since glacial times, with landslides a major factor in shaping its slopes.

Mirror Lake. Originally dammed by rockfall from adjacent cliffs, Mirror Lake grows smaller with each passing year. For a time the reflective lake surface was preserved by an artificial dam that raised the water level; the Park Service also for a time dredged the lake. Both damming and dredging are now frowned upon, and nature is allowed to take its course. Eventually the lake will fill completely with stream-carried sand, gravel, and clay, and with dead and dying plant material.

Nevada Fall. This fall, 181 meters (594 feet) high, plunges over a cliff whose position is governed by joints in the rock. Joints are widely spaced just above the fall, where the massive rock withstood glacial plucking, remaining as one of the treads in the glacial stairway of Little Yosemite Valley.

The terminal moraine of the Nevada Fall Glacier is just above the fall. Rock surfaces at the fall were smoothed by the earlier, longer glaciers. Here at the site of the present waterfall those glaciers must have broken and tumbled in an icefall of the type seen today on many mountain glaciers, such as those on Mount Rainier or Mount McKinley.

Nevada (upper) and Vernal falls tumble from steps of a glacial staircase. Massive granite forms the lip of each waterfall; weaker jointed granite below each waterfall was carried away by the ice.

Liberty Cap and Mount Broderick north of the fall are classic domes formed by glacial erosion and scaling off, exfoliation, along pressure-release joints.

Sentinel Dome (elevation 2476 meters, 8122 feet). A favorite sunset spot accessible by an easy walk, this dome is typical of many in Yosemite. It is rounded by a combination of glaciation (as indicated by patches of glacial polish and the many erratic boulders perched on its summit) and exfoliation along pressure-release joints. The vantage point offers good views north across Yosemite Valley, west to the Sierra foothills, and east and southeast to the High Sierra.

Taft Point and the Fissures. The short, pleasant walk from Glacier Point Road to Taft Point offers an opportunity to see, up close, joints in the granite widened by weathering, water, and wind. Nearby, Taft Point furnishes a view down the western part of Yosemite Valley, with El Capitan directly opposite and Cathedral Spires to the west.

Tenaya Lake. One of the prettiest lakes in the Sierra Nevada, Tenaya Lake lies in a glacially carved, oval basin among barren crags and slopes of Half Dome and Cathedral Peaks Granite. The Cathedral Peaks Granite is easily recognized: It contains unusually large crystals of feldspar that project as distinct knobs from the bare rock. Talus develops below cliffs and steep slopes where exfoliating granite breaks loose.

Water in Tenaya Lake comes from Cathedral and Polly Dome Lakes; discharge from Tenaya Lake flows down Tenaya Canyon to the east end of Yosemite Valley, where it joins the Merced River.

The Cathedral Peaks Granite contains large rectangular feldspar crystals. Resisting erosion to a greater degree than the granite matrix, some form handy grips and toe-holds for rock climbers.

Tioga Pass Road. Built as the "Great Sierra Wagon Road" to supply mines near Tioga Pass, this highway was purchased and repaired, as a gift to the people, by Stephen T. Mather in 1915. It gradually ascends the forested west-tilting slope of the Sierra to Tuolumne Meadows, and then climbs to Tioga Pass. Interpretive roadside displays point out many geologic features along the route. (See also Tenaya Lake and Tuolumne Meadows.)

At Tioga Pass, 3031 meters (9945 feet) high, the road crosses an eroded roof pendant of altered, metamorphosed sedimentary and volcanic rock—the rock that makes up nearby Mount Dana, second highest peak in the park. In places this rock is mineralized, but not enough for successful mining. There is a tiny ghost town, Bennettville, near some mines, about 3 kilometers (2 miles) by trail north of the pass—a worthwhile hike providing good views of alpine tundra and glimpses of Gaylor and Granite Lakes, as well as a beautiful semicircular cirque nestled below the divide. Far to the south is Mount Lyell, the park's highest summit, with a small, broad glacier in sight on its northern slope.

Tuolumne Meadows. This alpine meadow, largest in the Sierra, was buried in El Portal, Bridalveil, and Nevada Fall times by great ice caps that spawned glaciers flowing down Tenaya, Little Yosemite, and Tuolumne Valleys. The El Portal ice cap was more than 600 meters (2000 feet) thick over Tuolumne Meadows. Its slow outward flow scoured a shallow depression in the bedrock surface, a depression occupied much later by rock-studded ponds. As the ponds filled in with sand, gravel, and plant material, the meadow came into being. The Tuolumne River winds lazily from one end of the meadow to the other, its shifting course isolating oxbow lakes in abandoned meanders.

A classic cirque lies below the crest of the Sierra north of Tioga Pass. Granite Lakes are hidden within it. In the foreground is Gaylor Lake, dammed by a landslide.

Tuolumne Meadows occupy a broad valley shaped by the great icecaps of Pleistocene time. Mount Dana and Mount Gibbs lie beyond the head of the valley.

Lembert Dome at the east end of Tuolumne Meadows is a super-sized roche moutonnée. Such large granite bosses are glacially rounded on their upstream surfaces; downstream cliffs were steepened as moving ice plucked away great blocks of granite. Dark streaks are caused by weathering processes and, in part, by lichens.

Where the Tuolumne River rushes over granite ledges produced by pressure-release jointing, the water arches in foaming "waterwheels."

Many of these lakes are now filled in with lush plant growth.

Ice also rounded and polished numerous roches moutonnées, including the large granite domes that border the meadow, and left them strewn with erratics. Direction of ice flow is indicated by the lopsidedness of the roches moutonnées: The gentler slopes are the upstream sides, rasped and smoothed by the glaciers; the steeper downstream cliffs formed as moving ice plucked away chunks of rock. Glacial striae and polished rock surfaces are common here, especially on the large domes, and in places crescentic chattermarks show where ice-carried rocks ground against the underlying bedrock. The Tuolumne Glacier left the meadow from its northwest end, and flowed down the Tuolumne River's canyon to Hetch Hetchy Valley.

From different parts of Tuolumne Meadows the view includes several jagged peaks or horns that projected above the ice cap and so were never rounded off: Cathedral Peak, Unicorn Peak, and the Cockscomb to the south, and Ragged Peak to the north.

Ray Strauss photo

By late summer, Yosemite Falls may run out of water.

Vernal Fall. This is the lowest fall of the glacial staircase that extends upstream through Little Yosemite Valley. The position of the fall, governed by joints in otherwise massive Half Dome Granite, is at right angles to that of Nevada Fall. Today the rock continues to be undermined by frost and gradual decomposition of ever-wet rocks.

Wawona Tunnel Overlook. See **Yosemite Valley** and **Glacier Point.**

Yosemite Falls. Leaping 436 and 98 meters (1430 and 320 feet), Upper and Lower Yosemite Falls plunge from a high hanging valley nearly to Yosemite Valley's floor. Yosemite Creek, in the hanging valley above the falls, is fed largely by snowmelt, so the falls usually decrease to a trickle or appear to dry up completely by late summer.

The rock wall of this part of Yosemite Valley consists of strong, massive El Capitan and Sentinel Granites. The Upper and Lower Falls are separated by a zone of horizontal fractures which permitted more glacial plucking and which now continue to erode back from the base of the Upper Fall.

Yosemite Valley. Yosemite Valley tells an eloquent story of uplift and tilting of a major block of the Earth's crust, and of streams, rivers, and glaciers that cut into the uplifted block. The glacial trough branches upstream into Little Yosemite (Merced River) Valley and Tenaya Canyon. Its steep walls were shaped mostly by the El Portal Glacier, which was 1200 meters (4000 feet) thick where the Merced and Tenaya glaciers met.

The actual rock floor of Yosemite Valley lies well below its present grassy meadows. The terminal moraine of the Bridalveil Glacier, of which only the very top is now visible, dammed the valley, creating Lake Yosemite. As the lake filled with stream-carried sediments, the valley floor became almost flat.

For more on Yosemite Valley, see **Bridalveil** and **Yosemite Falls, Half Dome, Glacier Point,** and **El Capitan.**

OTHER READING

Arkley, R. J., Calkins, F. C., and others. 1962. *Geologic Guide to the Yosemite Valley, California.* California Division of Mines and Geology Bulletin 182.

Calkins, Frank. 1985. Bedrock Geologic Map of Yosemite Valley, Yosemite National Park, California. U.S. Geological Survey (with discussion of geology by N. King Huber and Julie A. Roller).

Ditton, R. P. and McHenry, D. E. 1976. *Yosemite Road Guide.* Yosemite Natural History Association.

Jones, W. R. 1971. *Yosemite, the Story behind the Scenery.* KC Publications, Las Vegas, Nevada.

Jones, W. R. 1976. *Domes, Cliffs and Waterfalls, a Brief Geology of Yosemite Valley.* Yosemite Natural History Association.

Jones, W. R. 1981. *Ten Trail Trips in Yosemite National Park.* Outbooks, Golden, Colorado.

Muir, John. 1912. *The Yosemite.* Reprinted in 1962 by American Museum of Natural History, New York.

Yosemite Valley's near-vertical walls are the upper walls of a deep glacial trough. Lake deposits cover the actual valley floor.

Glossary

aa (pronounced ah-ah)—a type of lava flow characterized by an extremely rough surface, with lava broken into irregular, clinkery fragments.

alaskite—granitic rock containing few or no dark minerals.

alluvial—deposited by rivers and streams.

alpine glacier—a glacier originating in mountains and flowing down valleys, also called a valley or mountain glacier.

andesite—dark gray volcanic rock composed largely of feldspar, often having large feldspar crystals in a finer matrix.

anticline—a fold that is convex upward.

arête—a narrow rock ridge sculptured by glaciers.

ash—fine particles of pulverized rock blown from a volcanic vent.

ashflow—volcanic ash that moves down the side of an erupting volcano rather than rising into the ash cloud.

badlands—rough, gullied topography in arid and semiarid regions, eroded by infrequent but heavy rains.

basalt—dark gray to black volcanic rock poor in silica and rich in iron and magnesium minerals.

basin—a downwarped, more or less equidimensional synclinal area whose youngest rocks are in the center; also any depressed, sediment-filled area.

batholith—a large mass of intrusive rock, more than 100 square kilometers (40 square miles) in surface exposure.

bedding—layering or stratification.

bedrock—solid rock exposed at or near the surface.

bentonite—soft, porous, light-colored rock formed by decomposition of volcanic ash.

biotite—black mica.

blowout—a depression or hollow formed by wind erosion, often found among sand dunes.

bomb, volcanic—a fragment of molten or semi-molten rock thrown from a volcano.

breadcrust bomb—a volcanic bomb with a cracked outer surface resulting from internal expansion after the crust cooled.

breakdown—large blocks of rock material fallen from cave walls and ceilings.

breccia (rhymes with "betcha")—volcanic rock consisting of coarse, broken rock fragments imbedded in finer material such as volcanic ash.

calcite—a common rock-forming mineral ($CaCo_3$), the principal mineral in limestone, marble, chalk, and travertine.

calcareous—made of or containing calcium carbonate.

calcium carbonate—$CaCO_3$, calcite.

caldera—a broad, basin-shaped volcanic depression formed by explosion or collapse of a magma chamber.

caliche (ca-LEE-chee)—calcium carbonate found on or near the surface of the soil in arid and semiarid climates.

cave onyx—banded calcite formed in a cave.

chert—a hard, dense form of silica that usually occurs as nodules in limestone.

chitin—a hard, protein-like substance common in skeletons of invertebrate animals.

cinder cone—a small, conical volcano built primarily of loose fragments of popcorn-like volcanic material thrown from a volcanic vent.

cinders—bubbly, popcorn-like volcanic material.

cirque—a steep-walled, usually semicircular basin excavated by the head of a glacier.

columnar jointing—a polygonal joint pattern, caused by shrinkage during cooling, that creates vertical columns in lava and volcanic ash.

conduit—the feeder pipe of a volcano.

conglomerate—rock composed of rounded, waterworn fragments of older rock.

crater—the funnel-shaped hollow at or near the top of a volcano, from which volcanic material is ejected.

creep—slow downhill movement of soil and rock.

crevasse—a deep fissure in a glacier, caused by movement over an uneven surface.

crystalline rocks—intrusive and metamorphic rocks whose crystals are large enough to be seen without magnification.

dacite—volcanic rock with a high proportion of quartz and feldspar.

dendritic drainage—a treelike pattern of branching streams and rivulets.

diatom—a microscopic, single-celled plant with a silica skeleton.

diatomite—rock formed of great numbers of diatom skeletons.

differential erosion—erosion at different rates regulated by differences in resistance of various rock types.

differential weathering—weathering at different rates regulated by differences in resistance of rock types.

dike—a sheetlike intrusion that cuts vertically or nearly vertically across other rock structures. In igneous rocks, dikes are often called **veins**.

diorite—a medium-gray igneous rock, the intrusive equivalent of andesite.

dome—an anticline in which rocks dip away in all directions. (See also **lava dome**.)

dripstone—travertine deposited by dripping water, as in stalactites and stalagmites.

earth flow—downslope movement of a well defined mass of soil and weathered rock, usually as a result of saturation with water.

epoch—a unit of geologic time, subdividing a period.

era—the largest unit of geologic time.

erratic—a boulder left by glaciers, commonly of a different rock type than nearby bedrock.

exfoliation—a process in which concentric sheets of rock break away from a rock surface.

extrusive rock—rock formed of magma which reaches the surface and solidifies there (also called **volcanic rock**).

fan—a cone-shaped mass of gravel and sand deposited by a stream, usually at an abrupt change in slope.

fault—a rock fracture along which displacement has occurred.

fault block—a segment of the Earth's crust bounded on two or more sides by faults.

fault scarp—a steep slope or cliff formed by movement along a fault.

fault zone—a zone of numerous small fractures that together make up a fault.

feldspar—a group of common, light-colored, rock-forming minerals containing aluminum oxides and silica. Feldspars constitute 60% of the Earth's crust.

finger lake—a long, narrow lake in a glacier-excavated valley, commonly dammed by a moraine.

flood basalt—flat-lying basalt that flooded across large expanses of terrain.

floodplain—relatively horizontal land adjacent to a river channel, with sand and gravel layers deposited by the river during floods.

flowstone—travertine deposited in caves by water trickling across cave walls and floor.

fold—a curve or bend in rock strata.

formation—a mappable unit of stratified rock.

fossil—remains or traces of a plant or animal preserved in rock; also long-preserved inorganic structures such as fossil ripple marks.

freeze-and-thaw weathering—prying apart of rock by crystal expansion as water freezes repeatedly in rock crevices. Also called **frost wedging.**

fumarole—a vent through which volcanic gases or vapors are emitted.

gabbro—a dark gray to black, crystalline igneous rock, the intrusive equivalent of basalt.

geophysical—pertaining to the use of instruments to measure physical properties of geologic structures.

glacier—a large mass of ice driven by its own weight to move slowly downslope or outward from a center.

globularite—a type of cave ornament that resembles popcorn.

gneiss—banded metamorphic rock thought to form from granite (which it commonly resembles), sandstone, and other continental rocks.

graded bedding—sedimentary layering with a gradual change in sediment size from coarse at the bottom to fine on top.

granite—a coarse-grained igneous intrusive rock composed of chunky crystals of quartz and feldspar peppered with dark biotite and/or hornblende. Also, in a broader sense, any light-colored, granular intrusive rock.

granodiorite—coarse-grained intrusive rock with less quartz and more feldspar than granite.

graywacke—a type of sandstone or conglomerate in which the grains are bits of sedimentary or volcanic rock such as slate or basalt.

greenstone—altered volcanic rock with a greenish color.

groundwater—subsurface water, as distinct from rivers, streams, seas, and lakes.

grus—disintegrated granite, its coarse, sandlike texture and angular grains derived directly from the parent rock.

halite—common salt (NaCl), a mineral formed by evaporation of seawater.

hanging valley—a glacial valley whose mouth is high up on the wall of a larger glacial trough.

harmonic tremor—undulating earthquake movement, fairly regular in pattern.

hematite—a common dark reddish brown iron oxide mineral, Fe_2O_3.

honeycomb weathering—weathering that creates numerous small, deep pits on a rock surface.

hornblende—a black or dark green mineral whose rodlike crystals are common in igneous rocks.

hot spring—a spring whose water temperature is higher than body temperature (37°C, or 98.6°F).

hydrothermal—having to do with hot water.

ice cap—a dome-shaped glacier covering the summit of a mountain mass.

icefall—part of a glacier deeply crevassed because of a steep drop in the valley floor beneath it.

ice field—a mass of ice, formed from compacted snow, that is not large enough or thick enough to move as a glacier.

ichthyosaur—an extinct group of swimming reptiles.

igneous rock—rock formed from molten magma.

ilmenite—a black mineral high in iron and titanium, $FeTiO_3$.

inclusion—a fragment of older rock included within igneous rock.

intrusive rock—igneous rock created as molten magma intrudes pre-existing rocks and cools without reaching the surface.

joint—a rock fracture along which no significant movement has taken place.

kettle—a steep-sided depression in a moraine or outwash plain, formed when a detached block of glacial ice melts.

lapilli—sand-sized to popcorn-sized fragments thrown from a volcano.

lateral moraine—a ridgelike mass of broken rock material deposited at the side of a glacier.

lava—molten magma that has reached the Earth's surface, or the rock formed when such magma cools.

lava dome—a type of volcano characterized by very thick magma that piles into a rounded dome above its conduit. Also called **volcanic dome.**

lava tunnel—a lava cave formed when fluid lava flows out from beneath its hardened crust.

lichen—a plant community consisting of a fungus and an alga, appearing as a flat, circular crust on a rock surface.

lime—a term commonly, though incorrectly, used for calcium carbonate.

limestone—a sedimentary rock consisting largely of calcium carbonate.

limonite—a yellow-brown iron oxide mineral ($2Fe_2O_3.3H_2O$).

lithified—turned to stone.

longshore current—a current that parallels a shore, usually established by waves striking the shore obliquely.

maar crater—a low-relief volcanic crater formed by an explosive steam eruption when hot magma contacts water.

magma—molten rock. When extruded onto the Earth's surface, magma is usually called **lava.**

magma chamber—a reservoir of magma from which volcanic materials are derived, usually only a few kilometers below the surface.

magnetite—a black, strongly magnetic iron mineral, $(Fe,Mg)Fe_2O_4$.

mantle—the zone between the Earth's core and crust.

marble—a metamorphic rock derived from limestone.

marine terrace—a nearly flat surface carved by waves and later elevated above the sea.

matterhorn—a glacier-sharpened peak, commonly with cirques on all sides.

medial moraine—a long moraine between two merging glaciers.

metamorphic rock—rock derived from pre-existing rocks as they are altered by heat, pressure, and other processes.

metasedimentary rock—sedimentary rock altered by heat, pressure, and other processes but still retaining some sedimentary characteristics.

metavolcanic rock—volcanic rock altered by heat, pressure, and other processes but still retaining some volcanic characteristics.

mica—a group of complex silicate minerals characterized by shiny, closely spaced, parallel layers that can be split apart easily.

mica schist—schist containing a large proportion of mica, which gives it a silvery, lustrous appearance.

mineral—a naturally occurring inorganic substance with characteristic chemical composition and frequently with typical color, texture, and crystal form.

moraine—rock debris deposited by a glacier.

mudflow—a flowing mass of fine mud. If much other debris is present, may be called a **debris flow.**

mud pot—a type of hot spring containing an abundance of mud.

muscovite—white mica.

névé—beaded ice formed by recrystallization of snow.

normal fault—a fault in which the hanging (upper) wall moves downward relative to the footwall.

obsidian—black volcanic glass.

olivine—a green mineral common in mantle-derived basalt and gabbro.

opal—a silica mineral containing up to 20% water, often with an iridescent play of color.

oreodont—an extinct group of sheeplike mammals common in Tertiary time.

orthoclase—a mineral of the feldspar group.

outcrop—bedrock that appears at the surface.

outwash—stratified sand and gravel deposited by streams of meltwater draining the front of a glacier.

overthrust—a low-angle fault in which one part of the crust slides over another, placing older rock on top of younger; also used for the oversliding block.

oxbow lake—a lake formed in an abandoned meander.

pahoehoe (pronounced pa-HO-ay-HO-ay)—ropy, undulating lava, usually basalt.

Pangaea—a supercontinent composed of all the present continents joined together.

pegmatite—exceptionally coarse-grained igneous rock found as dikes or veins in large igneous intrusions.

period—a subdivision of geologic time shorter than an era, longer than an epoch.

perlite—volcanic glass cracked into tiny, beadlike spheres.

phenocryst—a large, conspicuous crystal in a matrix of finer-grained igneous rock.

phyllite—shiny metamorphic rock having abundant mica crystals, intermediate between slate and mica schist.

pillow lava—lava (usually basalt) with a pile-of-pillows appearance characteristic of underwater eruptions.

plagioclase—a mineral of the feldspar group.

plate—a block of the Earth's crust, separated from other blocks by mid-ocean ridges, trenches, and/or collision zones.

plateau—a flat-topped tableland more extensive than a mesa.

Plate Tectonic Theory—a theory that states that sea floors spread and continents move apart as new crust is created at mid-ocean ridges.

playa—a flat-floored, vegetation-free lakebed that dries up quickly after rains, characteristic of desert basins with no external drainage.

pluton—a single igneous intrusion.

pocket beach—a small crescent-shaped beach, usually bordered by cliffs.

porphyry—igneous rock that contains conspicuous large crystals (**phenocrysts**) in a fine-grained matrix.

pothole—a small but deep circular depression excavated by the grinding action of pebbles, cobbles, and sand swirled by running water.

pressure-release joint—a joint concentric with the surface of once-buried rock, forming by release of pressure as overburden is washed away.

pterosaur—an extinct flying reptile.

pumice—light-colored, frothy volcanic rock, often light enough to float on water.

pyrite—a common brass-yellow iron mineral, FeS_2, also known as "fool's gold."

pyroclastic—rock material fragmented by a volcanic explosion.

quartz—crystalline silica (SiO_2), a common rock-forming mineral.

quartzite—sandstone consisting chiefly of quartz grains welded so firmly that, when broken, the rock breaks through rather than around the grains.

quartz monzonite—a granitic rock containing a large proportion of certain potassium and sodium feldspar minerals.

residual soil—soil that develops in place as rock disintegrates.

reverse fault—a fault in which the hanging (overhanging) wall moves upward relative to the footwall.

rhyodacite—extrusive igneous rock intermediate between dacite and rhyolite.

rhyolite—light gray volcanic rock with large quartz and feldspar crystals in a finer groundmass, the fine-grained extrusive equivalent of granite.

ring fracture—an arcuate fracture, one of a set surrounding a volcano, often associated with caldera formation.

roche moutonnée—a glacier-rounded rock boss, with a gentle upstream slope and a steep downstream slope.

rock avalanche—a fast-moving downslope flow of rock fragments supported by a self-generated cushion of compressed air.

rock flour—finely ground rock material pulverized by a glacier.

rock glacier—a mass of angular boulders and other rock material with enough interstitial ice to lubricate slow downhill movement.

rockslide—a landslide involving a large proportion of rock.

roof pendant—a downward projection of previously existing rock into an igneous intrusion.

schist—crystalline metamorphic rock which splits easily along parallel planes, commonly formed from fine-grained sedimentary rock. See also **mica schist.**

scoria—very bubbly volcanic rock, darker and heavier than pumice and with larger bubble holes.

sea-floor spreading—movement of oceanic crust away from mid-ocean ridges by creation of new oceanic crust at the ridges.

sea stack—a small, steep-walled island isolated from the mainland by wave erosion.

sediment—fragmented rock, as well as shells and other animal and plant material, deposited by wind, water, or ice.

sedimentary rock—rock composed of particles of other rock transported and deposited by water, wind, or ice.

seismic—having to do with earthquakes or earth tremors.

seismometer—an instrument for measuring earthquakes.

serpentine—a metamorphic rock having a silky or greasy luster and a slightly soapy feel.

shale—fine-grained mudstone or claystone that splits easily along bedding planes.

shield volcano—a broadly dome-shaped volcano formed by moderately fluid lava.

silica—a hard, resistant mineral, SiO_2, which in its crystal form is quartz. It also occurs as opal, chalcedony, siliceous sinter, and chert.

siliceous sinter—white, lightweight, porous silica deposited by some hot springs and geysers.

sill—a flat igneous intrusion that pushes between layers of stratified rock.

slate—fine-grained metamorphic rock that splits along planes that are not the original bedding surfaces.

slump—a landslide in which rock and earth slide as a single mass along a curved slip surface.

soil creep—gradual downhill movement of soil and loose rock.

spatter cone—a small, steep-sided cone built by molten lava that spatters from a volcanic vent.

speleothem—a cave ornament.

spit—a long, low point of beach sand, extending seaward or parallel to the shore, with one end attached to the mainland.

stalactite—a cylindrical or conical cave ornament hanging from a cave ceiling.

stalagmite—a cylindrical or columnar cave ornament projecting upward from the floor of a cave.

strata—layers of sedimentary (and sometimes volcanic) rocks.

stratified—layered.

stratovolcano—a volcanic mountain built of alternating layers of lava, breccia, and volcanic ash.

striae—parallel scratches created on rock surfaces as glaciers grind rock against rock.

subduction—the downward plunge of an oceanic plate below a continental plate.

swash mark—the arcuate line of large sand grains, mica flakes, seaweed, or other material marking the farthest advance of a wave up a beach.

syncline—a fold that is convex downward.

talus—a mass of large rock fragments lying at the base of the cliff or steep slope from which they have broken.

tarn—a small, deep lake occupying an ice-gouged basin.

tephra—broken rock material thrown out by a volcanic explosion.

terminal moraine—an arc-shaped moraine that marks the farthest advance of a glacier.

thrust fault—a low-angle fault on which older rocks slide over younger ones.

titanothere—an extinct mammal of Tertiary time, resembling a rhinoceros.

travertine—a hard, dense limestone deposited by lime-laden water of streams, hot springs, and caves.

tuff—rock formed from volcanic ash.

turbidity current—a downslope current caused by the weight of fine rock material suspended in water.

type locality—the locality at which a formation or other rock unit is well exposed and described as a standard for comparison.

unconformity—a substantial break or gap in the geologic record, caused by an interruption of deposition, by uplift and erosion, or by igneous activity.

valley fill—sand, gravel, and other rock material filling a valley.

valley glacier—see **alpine glacier**.

vein—a thin, sheetlike intrusion into a crevice, often with associated mineral deposits.

vent—any opening through which volcanic material is ejected.

vesicle—a small hole formed by entrapment of a gas bubble in cooling lava.

viscous—thick and sticky, not flowing easily.

volcanic dome—see **lava dome**.

volcanic plug—solidified lava that cooled within a volcanic conduit, generally more resistant than surrounding rock.

volcanic rock—rock formed from magma extruded and cooling on the surface.

welded tuff—rock formed of volcanic ash fused by its own heat, the heat of volcanic gases ejected with it, and the weight of overlying material.

Index

Page numbers in **boldface** indicate major discussions. Page numbers preceded by a "C" refer to the color section.

Adams, Mount, 87, 93
Agricola, Georgius, 1
Alaska, 9, 99, 111
Alberta, 40
Aleutian Islands, 9
Alta Vista, 87
Amphitheatre Point, 140, 141
Anacapa Island, 35, 37, 38, 39
Anderson, Mount, 120
Angeles, Mount, 118
Annie Spring, 44
anticlines, **17**, 31, 38, 109
Ape Creek, 97
Appalachian Mountains, 138
Applegate Formation, 124
arches, 26, 38, 39, 112–113, 131, 134
Arches, Point of, 114, 118
arêtes, 75, 89, 105, 136, 142, 144
ash flow tuff, (see tuff, ash flow)
Ash Mountain, 140
ash, volcanic, 39, 40, 41, 42, 55–59, 62, 75, 79, 90, **92–95**, C-1 (see also tuff)
Atlantic Basin, 105, 110, 131
Atlantic Ocean, 6, 16, 57, 105
Auburn, 89
avalanche chutes, 107, 139, 142, 144
avalanches, rock, 75, 81, 82, 89, 94, 96; snow, 26, 80, 91, 94, 107, 118

Backbone Ridge, 80, 87
badlands, 54
Bailey, Mount, 45
Baker Lake, 105
Baker, Mount, 105
Bald Crater, 47
Bald Hills Road, **132**
Bald Peak, 41
Bald Rock, 81
Baldy, Big, 136, 139
Baldy, Little, 136, 139
Barrier Peak, 81
basalt, **5**, 17, **72–74;** columnar (see joints, columnar); Columbia, 16, 59, 63, olivine, **5**, 52; pillow (see lava, pillow)
Basin and Range Province, 54
basins, 17
Basket Dome, 154
beaches, 27, 34, 35, 37, 112–113, 129, 132–134 (see also coastal processes)
Bearpaw Butte, 72, 75
bears, fossil, 59
Beaver Creek, Big, 105
Beaver Creek, Little, 105
Beetle Rock, 136, **138–139**
Big Arroyo, 142
Big Crack, 75
birds, fossil, 15
Black Crater, 74, 75
block diagrams, **27**
blowouts, 37
Blue Glacier, 111
Blue Mountains, 57
Bogachiel Peak, 120
Boiling Springs Lake, 71
bombs, volcanic, 43, 49, 62, 67
Boston Glacier, 107
Boulder Creek, 117
Bridalveil Falls, 147, 151, **152–153**
Bridalveil Glacier, 151, 153, 159
Bridalveil Granite, 152
Bridalveil Meadow, 153
British Columbia, 40, 74, 93
Broderick, Mount, 155
Brokeoff Mountain, 61, **64**, 70, C-2
Bubbs Creek, 144
Buckindy, Mount, 107
Buckley, 89
Buck Peak, 144
Bumpass Hell, 65, **66**
Bumpass Mountain, 63
Butte Lake, 66, 70
buttresses, 52, 53

Cabrillo Formation, 33
Cabrillo, Juan Rodriguez, 31
Cabrillo National Monument, **31–39**
Calawah Fault Zone, 117
calderas, **40–49**, **64**
Caldwell Butte, 72, 75
calendar, geologic, 12
caliche, 37
Cambrian Period, **12**, 13, 15
camels, fossil, 59
Canada, 12, 40, 111, 114, 117, 119
canyons, submarine, 32
Carbon Glacier, **80**
Carbon River, 80
Carrie Glacier, 118
Carrie, Mount, 108, 118
Cascade Creek (North Cascades), 106–107
Cascade Creek (Sequoia), 140
Cascade Pass, **106–107**
Cascade Range, 8, 40–49, 59–71, 75–90, 90–99, 100–107, 111
Cascade River Road, **106–107**
Castle Rock, 44
castles, 74, 75
Cathedral Lake, 155
Cathedral Peaks, 146, 157
Cathedral Peaks Granite, 155
Cathedral Rock (John Day), 57, 59
Cathedral Rocks (Yosemite), 153
Cathedral Spires, 155
cats, fossil, 59
Cave Creek, 121
cave ornamentation, 121
caves and caverns, 35, **121–124**, 128, 138, 139, 144
Cedar Grove, 142, 143, 144
Cenozoic Era, 12, 13, 14, 15, 39, **41–44**, **53**, 54, 55, **57–59**, **63–64**, **74–75**, **78–80**, **96–99**, **105–106**, **114–116**, **124**, 131, 138, 150–152
Central Valley, 135, 138, 144, 148, 150
Chagoopa Plateau, 142
Chalone Creek Fault, 126–128
Chalone Peaks, 125, 128
Channel Islands National Park, **35–39**
Chaos Crags, 62, 63, **66**, 70, C-2
Chaos Jumbles, **66**, C-2
Chavall, Mount, 107
Chelan, Lake, 107
Chilliwack Creek, 105
Cinder Butte, 75
Cinder Cone, 60–62, 63, **66–67**, 85
cinder cones, 17, 46, 47, 49, 60, 62, 63, 66–67, 72, 75, 85
cirques, **24**, 79, 80, 83, 89, 100, 105, 114, 136, 142, 144, 156
clams, fossil, 15
Clarno Formation, 58
Clarno Unit (John Day), 55
Cleetwood Cove, 44
climate, 15, 24, 54, 55, 58, 59, 63, 79, 114, 129, 131, 151, 154
coastal processes, **26–27**, **31–35**, 38, 39, **112–113**, **118–120**, **129–134**
Coastal Trail, 132
Coast Mountains (British Columbia), 40
Coast Range, 58, 74, 129–134, 154
coast redwoods, (see *Sequoia sempervirens*)
Cockscomb, 157
Columbia basalts (see basalts, Columbia)
Columbia Plateau, 16, 54, 55, 59, 63, 97
Columbia River, 95, 97
columnar jointing (see joints, columnar)
Conard, Mount, 61, 64
Conness, Mount, 152
continental drift, **3–9**, 34, 102, 105, 120, 124, 131, 150
corals, fossil, 15
core, Earth's, **5**
Coronado, 31
Cowlitz Box Canyon, **80**, 87
Cowlitz Chimneys, 81
Cowlitz River, 88, 95
crabs, fossil, 15
Crater Butte, 63
Crater Lake, 20, **40–49**, 64, 79, 88, 99, C-1
Crater Lake boat trip, **44**, 49
Crater Lake National Park, **40–49**, 61, C-1
creep, soil, 24, 25, 112, 118, 120, 141
creodonts, fossil, 58
Crescent Butte (Lassen), 69
Crescent Butte (Lava Beds), 72, 75
Crescent City, 131
Crescent Crater, 67
Crescent, Lake, **118**
Crescent Meadow, **139–140**
Cretaceous Period, **12**, 14, 31, 32, 35, 39, 53, 74, 105, 107, 128, 131, 138
crocodiles, fossil, 15
cross sections, **27**
crust, continental, 3, 6

crust, Earth's, 2, 3, **5,** 6, 8
crust, oceanic, 3, 6, 9
Crystal Cave, **139,** 144
crystalline rocks, **11**

Dana, Mount, 156
dating techniques, **13**
Denali, Mount, 155
Desert Cone, 41, 47
Devastated Area, **67–69,** 70
Devils Backbone, **44–45**
Devils Kitchen, 71
Devils Postpile Basalt, 53
Devils Postpile National Monument, **49–53**
Devonian Period, **12,** 13
Diablo Lake, 105, 107
Diamond Peak, 61, 64, 70
diatoms, 39, 67, 128
dikes, 40, 47, 48, 57, 79, 90, 136, 138, 145
Diller, Mount, 61, 64, 70
dinosaurs, 15
dogs, fossil, 59
domes, **17;** lava, **18,** 20, 24, 46, 52, **59–62, 69, 95,** 97, 99, 125, 126, 128
Dosewallips River, **116**
Double Peak, 81
drainage, dendritic, 99; radial, 63
Drakesbad, 71
dripstone, 122–124
dunes, sand, 26, 37, 133
Dutton Cliffs, 44

Eagle Nest Butte, 75
Eagle Peak, 63
Eagle Point, 118
Earth, age of, 1
earthflows, 26
earthquakes, 2, 6, 7, 37, 43, 49, 64, 75, 90, 91, 93–96, 99, 138
East Fork (Quinault River), 120
East Pacific Plate, **8,** 16, 17, 32, 41, 58, 93, 96, 105, 110, 111, 114, 124, 130, 138, 150
East Pacific Rise, **8,** 120
East Side Highway, **81**
El Capitan, 147, 148, 151, **153,** 155
El Capitan Granite, 148, **153,** 154, 158
Eldorado Peak, 107
Elijah, Mount, 121
Elk Prairie, **133**
El Portal, 151, 154, 155
El Portal Glacier, 151, 152–154, 156, 159
Elwha River, 115, **116–117,** 118
Emigrant Pass, 67, 70
Emmons Glacier, **81,** 82, 89
Emmons Glacier Vista, 89
Enchanted Valley, 120
Enumclaw, 89
Eocene Epoch, **12,** 39, 109, 114
erosion, **21–27;** coastal (see coastal processes); glacial, 22–23, 45 (see also individual glacial-erosion features)
erratics, glacial, 22, 63, 114, 136, 145, 146, 154, 157
eruptions, volcanic, **17–21, 40–44, 60–63, 72–75, 76–80, 90–99**
evolution, 1, 2, 15
explosions, steam, 66, 85; volcanic (see eruptions)
extinction, mass, 15
extrusive rocks (see volcanic rocks)

Fairfield Peak, 63
Fantastic Lava Flow, **66–67**
Farallon Plate, 9
Farview Dome, **147**
faults, 2, 3, 7, **15, 17,** 32–39, 53, 56, 72, 90, 96, 101, 110, 130, 131, 134, 136, 137, 142
fault zones, **17,** 81, 105, 117
Fern Canyon, **133**
Fern Creek, 131, 132
Fifes Peak Formation, 79
Fishes, Age of, **15**
fish, fossil, 15
Fissures, **155**
Flat Creek, 107
Flatiron Ridge, 71
Fleener Chimneys, 74, 75
floods, 117, 129, 134
Florissant Fossil Beds National Monument, 131
flowstone, 122–124, 139
folds, 2, 16, **17,** 109, 110 (see also anticlines)
Forest Lake, 64
fossils, 2, **15** (see also plant and animal names)
Franciscan Series, 129–134
Fremont, Mount, 89
Freshwater Lagoon, 134
frost, effects of, 21, 22, 25, 26, 101, 103, 105, 145, 150, 158
fumaroles, 46–47, 65, 66, 69, 70, C-2

Garfield Peak, **45**
Garibaldi, Mount, 40
Gavilan Range, 125
Gaylor Lake, 156
Generals Highway, **140–142,** 144
geologic time (see time, geologic)
George, Lake, **81,** 82
Giant Forest, 136, 138, 140
Gibbs, Mount, 156
Gibralter Rock, **82**
Gillems Bluff, 72, 74
Glacier Basin, 90
Glacier National Park, 105–111
Glacier Overlook, 89
Glacier Peak, 105
Glacier Point, 151, **153–154**
Glacier Point Road, 155
glaciers, 15, **22–23;** alpine, **24,** 49–53, 64, 75, **76–77, 79,** 89, 90, **100–103,105–107, 108–120,** 114–120, 136, 138, 142, **146–158;** continental, 105, 111, 114, 117, 118, 138
Glacier Vista Trail, 86
Glass Mountain, 75

Goat Rocks, 99
Gobblers Knob, **82**
Goblins Gate, 115, 117
gold, 100, 106, 107, 113, 133
Gold Bluffs, 131, 133, C-6
Gold Bluffs Beach, 131, 133, C-6
Gold Bluffs Formation, 13, C-6
Goodell Creek, 105
Grand Canyon of the Elwha River, 117
Grand Canyon of the Tuolomne River, 154
Grand Teton National Park, 111
Granite Lake, 156
granitic rocks, **9**
Grant Grove, 140–142, 144
gravity, effects of, 22, 26
"great hot blast", 59, 68, **69,** 70
Great Sierra Wagon Road, 156
Great Western Divide, 136, 137, 142, 144
Grizzly Peak, 153, 154
groundwater, 122

Half Dome, cover, 147, 149, 151, 153, **154,** 155, C-8
Half Dome Granite, 148, **154,** 158
hanging valleys, 75, 105, 106, 107, 120, 136, 144, 146, 148, 152, 154, 158, C-8
Harkness, Mount, 63
Hat Creek, 67
Hat Mountain, 63
Helen, Lake, **69**
Helen Mountain, 63
Hetch Hetchy Valley, 148, **154,** 157, C-8
High Sierra, 144, 146, 147, 154–155
High Sierra Trail, **142**
Hillman Peak, **45**
Hiouchi, 134
Hippo Butte, 72, 75
Hitchcock Lakes, 144
Hoh River, **117–118**
Holocene Epoch, **12**
Hood Canal, 114
Hood, Mount, 40, 76, 87
horses, fossil, 58–59
Horseshoe Basin, 107
Hospital Rock (Lava Beds), 74
Hospital Rock (Sequoia), 140, 141
hot spots, 8
hot springs (see springs, hot)
Hot Springs (Mount Rainier), **86**
Hot Springs Creek (Lassen), 63, 71
Hurricane Ridge, 109, **118**
Ice Ages (see Pleistocene Epoch)
ice caves, 87
ice, effects of, 24, 39, **76–77**
icefalls, 24, 76
ichthyosaurs, 15
Idaho, 95
igneous rocks, **9–11** (see also intrusive and volcanic rocks)
Indian Creek, 118
Ingraham Glacier, 81
Inspiration Glacier, 107
intrusive rocks, **9**

Ipsut Creek, 80
Island Butte, 75

John Day Basin, 56
John Day Formation, 55, 56, 58
John Day Fossil Beds National Monument, **54–59**
John Day River, 55, 57
John Muir National Monument, 131
joints, **17,** 35, 49–53, 90, 107, 122, 135, 136, 139, 140, 142, 145, 148, 154, 155; columnar, **9, 49–53,** 83, 86, 88; pressure-release, **21,** 135, 136, 138, 139, 141, 144, 145, 148, 149, 153–155, 157
Juan de Fuca Plate, 9, 105, 120
Juan de Fuca, Strait of, 114, 118
Juniper Butte, 74
Jurassic Period, **12,** 14, 32, 35, 39, 53, 105, 124, 131, 138, 150

Katmai, Mount, 99
Kautz Creek Mudflow, **81**
Kaweah Gap, 142
Kaweah Peaks Ridge, 142
Kaweah River, 136, 139, 140, 141, 142, 144, C-7
kelp forests, 34, 37, 39
Kent, Washington, 89
Kern Canyon, **142**
Kern River, 136, 137, 138, 142, 143, 144
Kern River Basin, 144
Kerr Notch, 45
Kings Canyon, **142–144**
Kings Canyon National Park, 131, **135–145,** 154
Kings River, 136, 142, 144
Klamath, 134
Klamath Marsh, 134
Klamath Mountains, 74
Klamath River, 129–133, **134**

La Jolla, 32, 34
La Jolla submarine canyon, 32
La Jolla Terrace, 32
Lake Chelan National Recreation Area, 107
landslides, 26, 34, 43, 54, 57, 59, 93, 112, 118, 120, 141; submarine, 31
Lassen Peak, 16, 40, 52, 59–71, 85, 128, C-2
Lassen Volcanic National Park, 16, 17, 21, 49, 52, **59–71,** 128, C-2
Lava Beds National Monument, 17, 49, **71–75**
lava domes (see domes, lava)
lava flows, 48, **49–53,** 57, 62, 67, 72, 74, 75, 77, 81, 82, 85, 87, 97–99, C-1
lava, pillow, **9,** 38, 39, 109, 112, 114, 116, 118
lava tubes, 71, **73,** 75
lava types of, **72–73**
Leaning Tower Granite, 152
leaves, fossil, 54, 55, 58, 131
Lembert Dome, 157
Liberty Cap, 153, 154, 155
lichens, 21, 74, 139, 144, 157
life, origin of, **15**
Linda Vista Terrace, 32
Little Ice Age, 152
Little Tahoma Peak, 81, 82
Little Yosemite Valley, 152, 153, **154,** 155, 156, 158, 159
Llao Rock, **45,** 46
lobsters, fossil, 15
Lodgepole, 141, 145
Lone Pine, 138
Longmire, **82,** C-3
Los Angeles, 39, 127, 136
Lost Creek, 67, 69, 70
Lyell, Mount, 152, 156
Lyre River, 118

magma, 8, 9
magnetism, Earth's, 13
Mammals, Age of, **15**
mammals, fossil, 15, 55–58
mammoth, fossil, 37
Mammoth Crater, 75
Mammoth Lakes, 138
Mammoth Mountain, 52, 53
Mammoth Pass, 53
man, evolution of, 1, 15
mantle, **5,** 6, 110, 134
Manzanita Lake, 69, **70**
map of region, 30
maps, geologic, 2, **27,** 102
Marble Creek (North Cascades), 107
Marble Falls, 139, 140
Marble Fork (Sequoia), 138, 139, 141, 144, 145, C-7
marine terraces, 27, 34, 37, 38, 39, 112, 116, 118, 131
Mariposa Grove, **154**
Maryland, 94
Mascall Formation, 55
Mather, Stephen, 156
matterhorns, 75, 100, 105, 136
Mazama, Mount, 20, 21, 40–49, 79, 87, 99
McLoughlin, Mount, 45
Medicine Lake Volcano, 72
Merced River, 148, 151, 153, 155
Merced River Canyon, 151, 152, **154–155,** 159
Mesozoic Era, **12,** 13, 14, 15, 32, **39, 53,** 74, **105, 124, 131, 138, 150**
metamorphic rocks, **101–11,** 14
metasedimentary rocks, **11**
Metasequoia glyptostroboides, 58, 138, 154
Middle Fork (Kaweah River), 140, 142, 144
Middle Fork (San Joaquin River), 49, 52, 53
mid-ocean ridges, 6, 7, 8
minerals, **11**
Miocene Epoch, **12,** 39, 74, 79, 87, 114, 127, 128
Mirror Lake, **155**
Mission Bay, 32
Mississippian Period, **12,** 13
Mist Falls, **144**
Modoc Plateau, 16, 71, 74
monitoring of volcanoes, 64, **79–80, 91–96,** 99
Mono Craters, 53
Montana, 95
moraines, 23, 24, 63, 64, 75, 81, 85, 86, 87, 89, 90, 100, 107, 118, 120, 136, 144, 145, 147, 151, 154–155
Moro Rock, 136, 137, 138, 139, 140, **144,** 145
Moro Rock Granite, 139, 144
mountains, formation of, **16–21**
Mount Rainier National Park, cover, 16, 17, 18, 61, **75–90,** C-3
Mount St. Helens National Volcanic Monument, 9, 16, 18, 20, 21, **90–99,** C-4
Muddy River, 95, 98
mudflows, 22, 55, 57, 58 59, 67, 68, 69, 70, 75, 77, 78, 80, 81, 84, **89–90, 90–99,** 128
mudpots, 59, 65, 66, 70
Muir, Mount, 144
Mushpot Cave, 73

Narada Falls, **85**
Nevada, 138
Nevada Falls, **155,** 158
Nevada Falls Glacier, 151, 152, 155
névé, 76
Newhalem, 107
New York, 94
Nisqually Glacier, 77, **85–86,** 87
Nisqually River, 82
Nisqually Valley, 86
Nisqually Vista, 86, 87
North American Plate, **8,** 16, 34, 41, 96, 105, 110, 114, 120, 130, 131, 138, 150
North Cascades Highway, 101, **107**
North Cascades National Park, 9, 14, 16, **100–107**
North Dakota, 40
North Dome, 144, 151, 153, 154
North Fork (Cascade River), 106
North Island, 31

Observation Point, 117
Obstruction Peak, 118
Ohanapecosh Formation, 79, 80, 81, 82, 86, 87, 89, C-3
Ohanapecosh River, **86**
Oligocene Epoch, **12,** 39, 88
Olympic Hot Springs, 117
Olympic Mountains, vi, 16, **108–120**
Olympic National Park, **108–120,** C-5
Olympic Peninsula, 108–120, C-5
Olympus, Mount, 108, 111, 118
Ordovician Period, **12,** 13, 105
Oregon Caves National Monument, **121–124**
oreodonts, fossil, 58
Osceola mudflow, **89–90**
outwash, glacial, 23–24, 118
Owens Valley, 136, **144**
oxbow lakes, 144
oxidation, 62, 74, 133, 140

Pacific Basin, 8

Pacific Coast, **31–34, 35–39, 108–120, 129–134,** C-5
Pacific Ocean, 5, 62, 105, 108
Painted Dunes, 67, C-2
Painted Hills Unit (John Day), 55
Paleocene Epoch, **12,** 39
Paleozoic Era, **12,** 13–15, **102–105,** 107, 124, **138, 150,** 154
Palisades, **86**
Panorama Point, 86, 87
Panther Creek, 105
Paradise, 85, **87**
Paradise Glacier, 87
Paradise Lost, 122
Paradise Ridge, 86, 140, 144
Paradise River, 85
peccaries, fossil, 59
pegmatite, 139, 145
Peninsular Ranges, 32, 34, 38, 39
Pennsylvanian Period, **12,** 13, 128
Permian Period, 12, 14, 128
petrified wood, 55, 131
petroglyphs, 73
Petroglyph Section (Lava Beds National Monument), 74
petroleum, 38, 39
Phantom Ship, 41, **46,** C-1
phenocrysts, **9,** 144
Picket Range, 103
Picture Gorge Basalt, 55, 56, 57
petroglyphs, 73
pigs, fossil, 58
pillow basalt (see basalt, pillow)
Pilot Pinnacle, 61, 64, 70
Pinnacle Peak, 89
Pinnacles, **46–47**
Pinnacles Fault, 126, 128
Pinnacles Formation, 127
Pinnacles National Monument, **125–128**
plants, effects of, 21, 22, 132, 133–134
plants, fossil, 15, 37, 54, 55, 58, 59, 131, 133
plateaus, 15
Plate Tectonic Theory, 1, 5, **6–7,** 16, 90, 110, 130
Pleistocene Epoch, **12,** 39, 49, 53, 59, 64, 74–76, **78–79,** 80, 82, 85, **105,** 111, 114, 116, 118, 124, 135, 136, 142, 146, 147, 151, 153, 156
Pliocene Epoch, **12,** 39, 105, 114
Point Conception, 37, 38
Point Loma, 31–34
Point Loma Formation, 34
Point of Arches, 114, 118
Polly Dome Lake, 155
Pothole Dome, 146
Poway Terrace, 32
Precambrian Era, **12,** 14, 105, 107
Pre-Lassen Dacite, 63
pressure-release joints (see joints, pressure-release)
Prospect Peak, 62, 63, 67, **70**
pterosaurs, 15
Puget Sound, 77, 80, 85, 111
Puget Sound Lowland, 82, 89, 105, 114, 118
pumice, 40, 43, 46, 49, 52, 53, 75, 79, 87, 90, 95, C-4
Puyallup, 89
Puyallup River, 82, 83, 89

Quail Flat, 141
Quaternary Period, **12,** 14
Queets River, **120**
Quinault Lake, **120**
Quinault River, **120**

rabbits, fossil, 59
Ragged Peak, 157
Rainbow Falls, 49, 52
Rainbow Falls Dacite, 52. 53
Rainier, Mount, 40, **75–90,** 93, 155, C-3
Raker Peak, 63, 67, 69, **70**
Rampart Ridge, 82
Rattlesnake Formation, 59
Reading Peak, 62, 63
Redcloud Cliff, 43, 44
Red Cone (Crater Lake), 41, **47**
Red Cones (Devils Postpile), 53
Reds Meadow, 53
Reds Meadow tuff, 53
Redwood Creek, 129, 131, 132, 134
Redwood Highway, **134**
Redwood Mountain, 141, 144
Redwood National Park, 13, **129–134,** 154, C-6
redwoods (see *Metasequoia, Sequoia, Sequoiadendron*)
Reptiles, Age of, **15**
rhinoceros, fossil, 59
Rica Canyon, 115, **116–117**
Rim Drive, 40, **47**
"Ring of Fire", 6, 9, 62
Ritter Range, 53
roches mountonnées, 105, 136, 146, 157
rock glaciers, 86, 87
rockslides, 32, 43, 54, 80, 85, 89, 96, 112, 128, 138, 153, 154 (see also avalanches, rock)
rock types, **10–11**
Rocky Mountain National Park, 112
Rocky Mountains, 57, 111
rodents, fossil, 58, 59
roof pendants, 53, 140, **144, 150,** 156
Rose Canyon, 32
Rose Canyon, 32
Rose Canyon Fault, 32
Ross Chimneys, 74, 75
Ross Lake, 105, 107
Round Pass, 82
Royal Arches, 153
Ruby Beach, 113
Ruby Creek, 107
Ruth, Mount, 81
Sacramento Valley, 150
San Andreas Fault, 3, 4, 8, 32, 35, 37, 125, 127–128, 138
sand dunes, 26, 37, 133
San Diego, 31, 32, 39
San Diego Bay, 31, 32, 34
San Francisco, 3
San Joaquin River, 49, 52, 53, 144
San Joaquin Valley, 150
San Luis Obispo County, 4
San Miguel Island, 37, 38, 39
San Nicholas Island, 37
Santa Barbara Channel, 38, 39
Santa Barbara Island, 37–38
Santa Cruz Island, 37, 38, 39
Santa Cruz Island Fault, 37, 38, 39
Santa Cruz Island Schist, 38
Santa Monica, 39
Santa Monica Mountains, 35, 38
Santa Rosa Island, 37, 38, 39
Santa Ynez Mountains, 38
Saskatchewan, 40, 43
Schonchin Butte, 72
Scott, Mount, **46**
sea caves, 26, 35, 37, 39, 112, 113, 129–134, C-6
sea cliffs, 33, 35, 37, 112, 118, 129–134, C-6
sea floor spreading, **8,** 105
sea stacks, 26, 35, 38, 39, 112–113, 118, 129, 131, 132, 134, C-5
Seattle, 108
seaweeds, effects of, 22
sedimentary rocks, **10–11,** 21
seismic monitoring, 64, 79–80, 91–96
Sentinel Dome, 144, 154, **155**
Sentinel Granite, 158
Sequoiadendron giganteum, 58, 131, 132, 135, 138, 154, C-7
Sequoia National Park, 71, 131, **135–145,** 154, C-7
Sequoia sempervirens, 131, 138, 154
Sespe Formation, 39
Seven Lakes Basin, **120**
Shannon Lake, 105
Shasta, Mount, 40, 45
Shastina, Mount, 45
Sheep Rock, 56
Sheep Rock Unit (John Day), 55, 57
shrinkage cracks, 49–63
Sierra Nevada, 6, 8, 16, 35, 38, 49, 53, 74, 111, 131, 133, **135–145, 145–157**
Sierra Nevada, 6, 8, 16, 35, 38, 49, 53, 74, 111, 131, 133, **135–145, 146–157**
Sierra Nevada Batholith, 32, 35, 39, 53, 135, 138, 142, 146, 148, 150
Silver Strand, 31, 32
sinter, siliceous, 66, 71
Siskiyou Range, 121
Skagit Gneiss, 14
Skagit River, 106, 107
Skagit Valley, 105
Skyline Trail, 86
Smith Creek (Mount St. Helens), 97
Smith River (Redwood), 131, **134**
snails, fossil, 15
soil, 21, 22, 69, 74, 79, 140
soil creep, 24
Soleduck Falls, 120
Soleduck Glacier, 120
Soleduck Hot Springs, 120
Soleduck River, 120
solution, 21, 22, 122, 139 (see also

caves, caverns)
Sourdough Ridge, 89
South Cowlitz Chimney, 81
South Dakota, 40
South Fork (Kings River), 142, 144
spatter cones, 17, 73–74, 75
spheroidal weathering (see weathering, spheroidal)
Spirit Lake, 90, 91, 93, 94, 95, 97
springs, hot, 41, 53, 59, 65, **66,** 70, 71, 86, 120, C-3; soda, 53
stalactites, 121, 124, 139
stalagmites, 121, 124, 139
Starr King, Mount, 153
Steamboat Prow, 81, 90
Steeple Rock, 118
Stehekin Valley, 105, **107**
Stevens Canyon Road, 80, 87
Stevens Peak, 89
Stevens Ridge Formation, 79, 80, 87, 89
St. Helens, Mount, 9, 16, 18, 20, 21, 40, 43, 47, 52, 54, 57, 62, 69, 75, 77, 79, 80, 85, 87, 88, **90–99,** 105, 128, C-4
strata, **11,** 27
stratified rocks, **11**
stratigraphic diagrams, **27**
stratovolcanoes, 17, 21, **40–49,** 61, 62, 63, 70, **75–90, 90–99,** 125, 128
striae, glacial, 40, 52, 63, 64, 65, 81, 85, 90, 100, 105, 107, 136, 142, 144, 152, 154, 157
Styx, River, 124
subduction, 8, 16, 32, 61, 97, 105 (see also Plate Tectonic Theory)
submarine fans, 32, 34
Sullivan, Lake, 118
Sulphur Works, 63, **70–71**
Sumner, 89
Sun Notch, 45
Sunrise, 81, **87–88**
Sunrise Ridge, **87–88**
Sunset Amphitheatre, 82
Sunset Crater, 85
Sunset Rock, **138–139**
Sur Series, 128
Survivors Hill, 70
Sylvia Falls, 87
synclines, **17**

Taft Point, **155**
Tahoma Creek, 82, 84
Tall Trees Trail, 132, 134
tapirs, fossil, 58
Tatoosh Pluton, 78, 79, 85, 87
Tatoosh Range, **89,** C-3
Tehama, Mount, **61–64,** 69, 70, C-2
temperature, effects of, 21, 22
Tenaya Canyon, 151–156, 159
Tenaya Creek, 153
Tenaya Lake, 152, **155,** 156
terraces, marine, **32,** 112, 131
Tertiary Period, **12,** 14, 35, 39, **78–79, 96–97,** 107, 114, 121, 138
Thielson, Mount, 45
Three Brothers, 150
Three Sisters, 72
Thunder Creek, 105
Tijuana River, 31
Timber Crater, 41, 47
time, geologic, **12–14**
Tioga Pass, 154, **156**
Tioga Pass Road, **156**
Tipsoo Lake, 81
titanotheres, fossil, 58
Tokopah Falls, **145**
Tokopah Valley, **145**
Toutle River, 69, 80, 93, 94–97
Toutle River mudflow, **94–96**
Trail Crest, 144
transportation of rock material, **21–23**
Transverse Ranges, 8, 35, 39
trenches, oceanic, 66, 105, 130 (see also Plate Tectonic Theory)
Triassic Period, 12, 14, 124, 150
Triple DIvide Peak, 142
tubes, lava, 71
Tueeulala Falls, 154, C-8
tuff, ash flow (welded), 57, 79, 80, 86, 87
Tule Lake, 72, 74, 144
Tunnel Falls, 140
Tuolumne Glacier, 157
Tuolumne Meadows, 148, 149, 154, **156–157**
Tuolumne River, 149, 154, 156, 157, C-8
turbidity current deposits, 34, 116
turtles, fossil, 15
type localities, 14

unconformities, 33
Unicorn Peak, 89, 146, 157
Union Peak, 41
United States Geological Survey, 27, 93, 96, 98, 99
Upper Soda Spring, 49, 53
upwelling (of ocean waters), 37
U-shaped valleys, 24, 40, 45, 63, 70, 75, 77, 79, 86, 88, 100, 101, 105, 107, 111, 120, 136, 137, 142, 144, 154

Valhalla, 142
Vancouver Island, 111
Vernal Falls, 155, **158**
Vesuvius, Mount, 76
volcanic rocks, **9–11, 17–21**
volcanoes, 2, 6, 9, 16, **17–21, 40–49, 49–53, 59–70, 71–74, 75–90, 90–99;** maar, 74; shield, 8, **62,** 67, 70, 72, 74
Volcanoes, Cascade, 17, **18–19,** 21, **40–49, 49–53, 59–70,** 72, 74, **75–90, 90–99,** C-1, C-2, C-3, C-4 (see also specific mountain names)
Vulcans Castle, 63, 64

Wapona Falls, C-8
Warner Mountains, 74
Warner Valley, 63, **70**
Washington Column, 153
Watchman, 44, **47–49**
Watchtower, 145
waves, effects of, **26–27**
Wawona Tunnel Overlook, 158
weathering, **21–22,** 27, 57, 118, 120
weathering, spheroidal, 136, 140–141
West Fork (Smith River), 131
White River, 81, 85, **89–90**
Whitney Butte, 72, 75
Whitney Crest, 136
Whitney, Mount, 135, 136, **144**
wind, effects of, 22, 24, 26, 57
Windy Ridge, 91
Winthrop Glacier, 81
Wizard Island, 20, 42, 43, 44, **49**
Wonderland Trail, **90**
wood, petrified, 55, 58
Woods Creek, 144
Wow, Mount, **82**
Wyoming, 40

Yakima Park, **87**
Yellowstone National Park, 21, 40, 112, 132
Yosemite Creek, 158
Yosemite Falls, **158,** C-8
Yosemite, Lake, 151, 152, 153, 159
Yosemite National Park, cover, 31, **146–159,** 71, C-8
Yosemite Valley, 142, 146, 147, 150, 151, 153, 155, **158–159,** C-8